POLARITIES IN THE EVOLUTION OF HUMANITY

POLARITIES IN THE EVOLUTION OF HUMANITY

WEST AND EAST—MATERIALISM AND MYSTICISM—KNOWLEDGE AND BELIEF

Eleven lectures given to members of the Anthroposophical Society in Stuttgart between 5 March and 22 November 1920

INTRODUCTION BY MATTHEW BARTON

RUDOLF STEINER

RUDOLF STEINER PRESS

CW 197

Rudolf Steiner Press
Hillside House, The Square
Forest Row, RH18 5ES

www.rudolfsteinerpress.com

Published by Rudolf Steiner Press 2022

Originally published in German under the title *Gegensätze in der Menschheitsentwickelung: West und Ost—Materialismus und Mystik—Wissen und Glauben* (volume 197 in the *Rudolf Steiner Gesamtausgabe* or Collected Works) by Rudolf Steiner Verlag, Dornach. Based on shorthand notes that were not reviewed or revised by the speaker. This authorized translation is based on the third German edition (1996), edited by Robert Friedenthal and Paul G. Bellmann

Published by permission of the Rudolf Steiner Nachlassverwaltung, Dornach

Translation revised by Matthew Barton

A catalogue record for this book is available from the British Library

ISBN 978 1 85584 601 2

Cover by Morgan Creative
Typeset by Symbiosys Technologies, Vishakapatnam, India
Printed and bound by 4Edge Ltd., Essex

Contents

Introduction

We tend to think of polarities in binary terms: the either/or of mind and matter, say, which both easily satisfies yet ultimately perplexes the merely rational mind since it does not find a way to bridge this and so many other divides. In the lectures collected in this volume, given in a Germany crushed and fragmented by the Great War and still suffering its aftermath, Steiner develops and elaborates ideas which, because they originate in a complex and living picture of human and social realities, can broaden narrower ways of thinking and challenge any merely schematic or programmatic template we might impose upon our experience of the world.

As I write these words, just over a hundred years since Steiner gave the lectures collected here, a long unhealed East-West divide between nations has erupted once more, with the potential to provoke global catastrophe. A condition of cold war we might think had been overcome has, it now becomes apparent, never been radically addressed. Mutually antagonistic national and economic interests have merely, throughout, been seeking to strengthen themselves at each other's expense, and Central Europe, instead of playing a dynamic, mediating role in the world, appears only as a co-opted satellite of these warring interests, seemingly unable to unfold an intrinsic power of its own. It is held, in Steiner's words'—albeit a century ago—in an imprisoning 'vice'. 'Things fall apart, the centre cannot hold', as W.B. Yeats wrote in his poem 'The Second Coming', published in the same year as these lectures were given.

One of the qualities that is so outstanding about Steiner, and vividly apparent here also, is his ability to connect socio-political realities with their source in human thinking, showing how the fixed divisions of our experience of the world become the sundered and warring social conditions we end up inhabiting. We make our own reality. And this reality, as he stresses here, is becoming ever less human, is succumbing increasingly to anti-human forces. Time and again he challenges the limitations of our hearts and minds, urging us to move beyond them into broader, more dynamic, truly human and therefore also healing perspectives.

Prophetically, Steiner identifies part of the (also self-inflicted) attack on human values and the less than human reality we are creating, in the rise of technology—which, as he says, embodies 'forces independent of humanity' that take hold of the unconscious human will, rising from there to impinge upon our conscious thinking and alienating us from a human evolution worthy of the name. 'Human beings have wholly surrendered themselves to something that is no longer human.' Yet with his always mobile and wide-ranging perspective, he does not seek to repudiate such 'necessary' developments, but urges us to find the sources of strength to 'integrate' them properly.

This inevitably involves us trying to become more conscious: 'We will not manage our evolution properly', he says, 'unless we make efforts to develop ... thoughts increasingly independent of mere feelings and emotions, of anything arising in us out of dreamlike inner experience even when we are fully awake. Theoretical principles will not help us achieve this, only life itself can do so... people are not prepared to listen to anything that does not arise from their own inner prejudices...'.

Holding fast to our own point of view, failing to attend empathically to the very different views of other people and nations, however misguided we may think them, or they may actually be, is to fan the flames of our own prejudices and lock us in to a divisive tribalism that runs counter to evolution; it is to sow discord and warfare within ourselves, firstly, and then by extension within the global reality we are creating. The antagonistic polarities of different extremes are one-sided ways of being and thinking that can ultimately be seen, in Steiner's view, as forms of sickness. To stop this endless seesaw we need to seek and hold the balance between all the outlooks and viewpoints, a healing, mediating and ultimately loving embrace of differences. We can try to 'speak with inner empathy about things we wish, nevertheless, to see overcome.' Truth is here seen not as a static position or a fixed possession but as a continual, active and loving deed.

This returns us continually, also, from the stage of global conflict to the soul's inner scenario and to our own responsibility for the world around us. Rather than passive victims of global forces, these forces issue from and can be transformed, albeit very slowly, by ourselves. In this evolutionary project nothing will be achieved either by a 'limitless belief in authority', in which we hand responsibility for our lives and wellbeing to 'experts', nor by a 'mystical small-talk' that gains no purchase on the current world situation. It becomes vividly apparent that only fluid and living concepts, not fixed terminologies or outlooks, can do justice to the dynamic flux of reality. Only an unwavering pursuit of the 'spirit of truth', embodied ultimately in the redemptive figure of Christ, will reach beyond all fixity and division, in ourselves and our thoughts and feelings firstly, and then also in society as a whole.

Yeats's great poem ends: 'And what rough beast, its hour come round at last, / Slouches towards Bethlehem to be born?' In Steiner's words, a spirit-numbing materialism is not untrue but is 'unfortunately becoming true'. If we are to retain our humanity against the odds, it is high time for us to awaken our thoughts from their brain-bound enchantment and learn to think differently.

Matthew Barton
March 2022

Lecture 1

STUTTGART, 5 MARCH 1920

THE challenges presented by our age really have to be faced by every individual person today. I have made it quite clear on a number of occasions that to understand the way individuals need to face those challenges we must be aware of how human evolution progresses everywhere in the world. The whole course of human evolution can only be clearly understood if we gain profounder insight into the powers that intervene in the course of earth evolution as a whole and also in human lives.

I have used a number of different approaches to show that as human beings we are part of an ongoing evolution that may be said to be taking its normal course. Spiritual science enables us to follow its progress over extended periods of time. I have also pointed out that there are certain powers with different goals for mankind than those that desire to guide humankind's normal course of evolution through the earth's various embodiments, powers we have been obliged to call luciferic on the one hand, and ahrimanic on the other. I have spoken of this a number of times. It is necessary to take a very serious view of these things today, but our hearts and minds cannot really achieve this serious mood unless we pay proper attention to the way these luciferic and ahrimanic powers intervene directly in human lives.

As you know, a new era in human evolution started during the fifteenth century, very different from anything that went before. Thinking of this you will want to be aware of the many ways life in the present era—which had its beginning in the fifteenth century—is different from the era preceding it. We may say that one particular feature of the present age is the development of intellectual thinking since the middle of the fifteenth century. In the great process of schooling that humankind has undergone in the course of earth evolution, it needed also to pass through this schooling of the intellect. Human beings had, as it were, to find out how life can be lived when intellectual thinking is accentuated. They could never have grown into truly free individuals without incorporation of the intellectual principle. We have no clear idea today of the extent to which people differed from us before the middle of the fifteenth century, particularly in this respect. We tend to take the things we are given for granted, without giving them much thought. We are now generally dealing with the peoples of civilized countries who are inclined to think with the intellect, and we have come to believe that people have always thought like this. That is not the case, however. Before the middle of the fifteenth century people thought in a different way. They simply did not think in the abstract terms in which we think today. Their thinking was very much more vivid and concrete, immediately bound up with the objects of the world around them. They were much more bound up with the feelings and will impulses that can be experienced in the human soul. We are living to a great degree in our thoughts, though we are not sufficiently aware of this. We are not even aware of the source from which this way of thinking, the intellectual approach which we take so much for granted, has evolved. We shall have to go a long way back in human evolution to get a real understanding of the origins of this way of thinking, this intellectualism. Another question we must ask ourselves is whether anything still remains of the human faculty from which our thinking has evolved.

You know that older evolutionary forces persist into later ages and continue to be present side by side with those that are normal to the era in question. This also applies to our thinking. Reminders, echoes of thinking, of an activity similar to our thinking, are experienced in our dreams, when a whole world of images emerges from our night-time sleep. Experience teaches us to distinguish between the world of thoughts we develop between waking up and going to sleep and the world of dream images which we experience in an entirely passive way. If we go back to earlier times in human evolution we find that the further back we go the more the life of the soul during waking hours comes to resemble the kind of soul activity we know in our dreams today. Present-day thinking is the fruit of later stages of evolution. During earlier stages along this path the human soul developed activities more akin to dreaming. If we trace this dreamlike activity of the human soul a long way back we find ourselves before earth evolution as we know it. We come to a time when the earth had taken a physical form in the cosmos that preceded the present one. We have become accustomed to calling this the stage of Old Moon evolution. Human beings were part of this as well, but in an entirely different form. During that Moon evolution, i.e. the time when the earth materialized in a form that preceded the present one, the human being, the true ancestor of modern humanity, was still completely etheric. His soul became active in a way that was definitely dreamlike, consisting of dream images. The peculiar thing about this was that it related to the outer world in a way that is quite different from the soul activity we know as thinking. I would say that when our soul is active in thought we find ourselves rather isolated within the world. The world is outside us, it has its own processes. We reflect on those processes in our minds, but precisely when we think we are reflecting most profoundly on these external processes, we actually feel ourselves entirely outside them. Indeed we often feel that we are best able to think about external phenomena if we keep ourselves well isolated from them, withdrawing into ourselves. The human ancestor who was dreamy in his thinking, if I may put it like this, did not have that feeling. Developing in his way in his dreams what we develop in our way when we are thinking, he knew himself to be intimately bound up with everything he experienced of the world. We see the clouds, we think about them, but we do not feel that the powers alive in the clouds are also alive in our thinking. Our human ancestor did have the feeling that the powers alive in a cloud were also alive in his thinking. This ancestor said—and I must translate what he said into our language, for his language was a silent one compared to ours: The powers that are alive and active in the cloud out there produce images in my mind. He saw himself no more isolated from the great universe in which the cloud revealed its essential nature than my little finger is able to think itself isolated from the rest of me. If I were to cut it off it would wither; it would no longer be my finger. The human ancestor felt that he could not exist apart from the universe that belonged to him. My little finger might well say: The blood which pulses through the whole of the body also pulses within me; the whole of my organic life is governed by the same laws as the organic life of the rest of the body. The human ancestor said: I am part of the universe; the power that pulses within me as I form images is the same as the power that is alive and active in the forming of clouds. That is how the human ancestor felt himself to be closely related, intimately bound up, with the whole world.

Because we must feel so isolated in our thinking from everything that goes on outside us, we are in a sense sundered and separate from the essential origins and causes of the existing world. In ordinary life we are not aware of the pulses beating throughout the universe. Our thinking has grown abstract. Our thinking tells us nothing, as it were, of what is alive and active within it. This actually provides the potential for human freedom, a freedom where we do not feel that something is thinking in us but that we ourselves do the thinking.

The ancestors of humanity were unable to form ideas independently of the universe as a whole. These forefathers of humanity felt themselves to be bound up with the existing world; they knew that this existing world contained more than just abstract forces of nature. They knew that beings, albeit ones different from human beings, held sway, beings that did not have a physical body such as the human body, though human beings might feel that they shared citizenry of the universe with them. These ancestors were not aware of the 'forces of nature'; they felt themselves to be in communion with nature spirits. Today we may say that everything that happens in nature follows the laws of nature, and we are part of that nature. For the human ancestor living in a far distant past it was natural to say that everything that happened in nature outside himself arose from the will impulses of nature spirits. We say the earth attracts the bodies that are on it due to gravity, and according to the law of gravity the gravitational pull decreases at a rate that is proportional to the square of the distance between the two. We call this a special case of a law of nature. Such abstract notions are the basis for our understanding of nature. The ancestors of humanity knew that an essential spiritual element was present in the phenomenon we have made into an abstract gravitational force.

Certain spiritual powers who may be said to be involved in human evolution thus developed a relationship to human beings that would normally cease the moment earth evolution proper began for the human being. At that point human beings would be released from the tutelage of those spiritual powers, powers they had felt to be flowing into and buoyant within them during the Old Moon stage. So we must ask ourselves what it was that made human beings grow independent of the guidance of spirits with whom they had felt at one, however dimly. It happened when the mineral kingdom became part of human nature. In those far distant times of which I have just spoken, human beings did not yet have the mineral kingdom within them. Their organization would not have been perceptible to our present sense organs, for it did not yet include mineral elements.

To grasp this without getting caught up in preconceived notions, we need to consider what it truly means when an organism incorporates the mineral kingdom. People tend to be superficial in their thinking about such things. We look at a mineral, a stone, and quite rightly consider it to be the way it presents itself to our observation. Then, however, we look at a plant in exactly the same way we look at a stone. In reality it is not the actual plant we see. A plant is really something entirely beyond sensory perception. Consider a system of forces that in a sense has the qualities of an image. Its relationship to the mineral kingdom is that this otherwise invisible organization absorbs the mineral kingdom and the forces that are active between individual constitutive elements in that kingdom. I have a plant before me. It is an invisible system of forces that absorbs mineral principles from the mineral kingdom. The result is that the mineral aspect occupies the space also occupied by the invisible system of forces. I see this mineral aspect, though it is merely something the supersensible plant has absorbed. I first have to discover the supersensible plant in a way quite different from how it appears to me after absorbing substance. This is already the case with a plant. When we talk about plants today we are really talking only of the minerals contained within them and not about the plants themselves. It is important that we clearly understand this in the case of a plant, for it also applies to animals and humans, only more so.

During the Old Moon stage, then, human beings did not incorporate mineral in this way. Human beings living on the present earth are so constituted that they need the mineral kingdom, have absorbed the mineral kingdom and its forces into them, as it were. What significance does this have for human nature? Above all human beings acquired a mineral body for their earlier pictorial thinking. As evolution progressed the mineral human body provided the basis for intellectual thinking that developed at a relatively late stage, from the middle of the fifteenth century onwards, having been a long time in preparation.

Modern intellectual thinking is based on the fact that a mineral body has been incorporated into human beings. As human beings we need a mineral body first and foremost in order to think. The older form of pictorial thinking had developed through what we call the third elemental kingdom. The mineral kingdom had the function of transforming this pre-earthly form of thinking into our earthly way of thinking. But because of this, the spirits with whom human beings inevitably felt connected, in their pictorial thinking in the distant past, were then relieved of their function. We will have to picture those spirits rather differently from the way we are accustomed to picture beings other than human ones. People, even people of goodwill who may admit that there is more to life than is apparent to the senses, tend to stick too closely to the human form. This anthropomorphism takes over whenever people try to create an image in their minds of anything that is higher than the human sphere. It is easy to accuse Feuerbach and Buechner[1] of anthropomorphism. We have seen more than enough of this kind of thing. We have seen the legal way of thinking evolve in the Western world, with earthly misdeeds and crimes judged by earthly judges who impose penalties, and so on. The supersensible realm has gradually come to be seen, in an imperfect form of Christianity, in terms of rewards and punishment meted out for sins, like the proceedings in an earthly court of law. The religious ideas of the West have a great deal of human jurisprudence in them. We let the gods mete out punishments of the kind we know earthly courts of law impose. If we truly wish to get beyond the merely human we must firmly decide not to think in entirely human terms. We must think beyond anything anthropomorphic, and that indeed is what really matters in human life. That is the approach we must use if we want to see clearly that the spirits who influenced human pictorial thinking during the Old Moon stage lost that function in the normal progress of human evolution but do not accept this with good grace. We might ask why they do not submit to the will of the gods who guide normal progress. They simply do not. We have to accept that as a fact. The original intention was that they should only influence dreams within the human sphere and everything related to dreaming. In the context of today's lecture we refer to them as luciferic spirits. Their proper sphere would be everything that has to do with dreaming and anything related to this. They are not satisfied with this, however. They infiltrate the form of human thinking that has evolved out of their own sphere, human thinking now bound to the mineral sphere. When we allow anything that normally rules our dreams, the life of fantasy, to enter into our thinking, we fall prey in our thinking to luciferic nature, to the influence of spirits that should only have influenced the old form of pictorial thinking that belonged to the ancestors of humanity. They have retained their power and instead of limiting themselves to our dreaming, our life of visionary fantasies, our creative artistic work, they are constantly trying to influence our thoughts and make them dependent on impulses similar to those that existed in pre-earthly times. Our thinking is still greatly influenced by elements coming from this source, by the luciferic principle.

It is justifiable to ask in all seriousness what powers these are that have such an influence on our thinking? These influences arise from the sphere where we human beings are still rightfully dreaming and rightfully asleep above all else. They come from the sphere of our feelings and emotions. We experience our feelings, after all, in the way we normally experience dreams and we experience our will the way we experience sleep. There we are still rightly cocooned in a world which becomes a luciferic world as soon as it develops in our thinking. We therefore will not manage our evolution as human beings properly unless we make the effort to develop other thoughts as well, thoughts increasingly independent of mere feelings and emotions, of anything arising in us out of dreamlike inner experience even when we are fully awake. Theoretical principles will not help us achieve this, only life itself can do so. We find, however, that the inner habits humankind has acquired put up great resistance to the cultivation of mind and soul that is needed. We must be on the lookout for this resistance. We find in the present time, in particular, that people are not prepared to listen to anything that does not arise from their own inner prejudices, their feeling of how things should go, their personal preferences. They are not in the habit of listening to anything which in a way has been decided independently of human beings, requiring merely their consent. I should like to give you a brief example which I used on one occasion to explain to someone an important difference concerning human thinking.

Many years ago I gave a lecture in a town in southern Germany—today no longer in southern Germany—on the wisdom taught in the Christian faith.[2] As you know, it is always necessary to limit the subject matter presented in a particular lecture and one can only speak within that context. When people hear just a single lecture, such a single lecture will impress one person in one way and another in a different way, particularly if you have been objective and dispassionate in presenting the subject. It certainly would not be possible for anyone to assume something about the overall outlook underlying a single lecture if they just listened to that one lecture. If the wisdom taught within the Christian faith is the subject for example, it will of course be impossible to conclude from the contents of the lecture what the speaker thinks about the connection between light and electricity, say. It is therefore possible for something to happen as it did on that occasion. I spoke about the tenets of Christian wisdom, and two Roman Catholic priests were in the audience. They came up to me afterwards and said: 'No objection can be raised to what you have been saying'—this, by the way, was many years ago now—'but we have to say that whilst it is true that we say the same thing, we say it in such a way that everybody can understand it.' My reply was: 'Reverend fathers, surely it is like this: You or I may have some kind of inner feeling that we are speaking for everybody, but that is not the point, for that is a subjective feeling. After all it is perfectly natural—if we go entirely by our feeling I, too, must believe that I am speaking for everybody, just as you think you do; that is self-evident; otherwise we would do it differently. But we are now living in an age when our belief that something is justifiable does not count. We need to let the facts speak for themselves. We must learn to look to the facts. Subjectively you believe you are speaking for everybody. But now let me ask you about the facts. Does everybody still come to your Church? That would show that you are speaking for everybody. You see, I speak to those who do not come to your Church to hear you speak. My words are for those who also have the right to hear of the wisdom taught in Christianity.' In this way we must allow what the facts tell us to govern us.

It is necessary for us to tear ourselves away from our subjective feelings. If we do not do so, the luciferic element will enter our thinking. We would not have had to suffer the truly dreadful campaign of untruthfulness that has become ubiquitous in the last five years, the final consequence of something that has long been in preparation, if people had learned to pay rightful attention to what the facts have to tell us rather than to their emotions, with nationalists the worst in stirring up such emotions.

On the one hand there is the absolute necessity today of educating our thinking to comply even with things that do not come naturally to us. Militating against this is people's dislike of a truthfulness that takes its lead from facts, from realities.

We shall not be able to attain higher worlds and the knowledge to be gained there if we do not strictly school ourselves through the realities of the external world. Having to some extent at least become accustomed to wanting to hear realities presented, you will often suffer tortures when people of the present age try to tell you something. Very often the kind of thing you hear people say is: 'Oh, someone said something that was frightful, quite terrible!' Terrible in what way? You say it was terrible but that only tells me how you felt about it. I really want to hear exactly *what* it was. 'Well, it really was terrible what was said there…' And these people simply do not understand. All the time they want to describe their subjective feelings concerning the matter, whilst you want to hear an objective report of what they actually witnessed. It is especially when people tell you something someone else has told them, that it is quite impossible to tell if they are simply passing on what they have heard or if they have actually investigated the matter they are talking about. This is an area where one has to remind people again and again that truthfulness concerning the knowledge to be found in supersensible spheres can only be achieved if we train ourselves as far as possible to adhere closely to facts in the sense-perceptible world. That is the only way in which human beings can overcome the luciferic elements that stream into their thoughts—by schooling themselves through perception of facts and realities.

On the one hand mankind is exposed to luciferic influences, on the other to ahrimanic influences. I said that thinking here on earth could only evolve from earlier stages of human soul life when human beings in a sense incorporated a mineral body. This mineral body is indeed the organ for the earthly way of thinking. It does however bring it predominantly into the sphere of the powers we call ahrimanic.

We can of course become aware of the need to school ourselves through a world of facts, so that we cease to be habitually swayed by our subjective emotions. We must not, however, fall prey to the kind of thinking that is nothing but an inner activity arising from the mineral body. Here we come upon a truth that many people find highly unpalatable.

You know how some are idealists or spiritualists and others are materialists. There is plenty of discussion in the world as to which is the right outlook, spiritualism or materialism. All these debates are of no value whatsoever for certain regions of the human organization. Human beings can develop in two ways. We can use the mineral body we have incorporated as the instrument for our thinking, and indeed we have to use it, otherwise we would merely be dreaming. But we can then also rise in our thinking beyond this instrument; we can develop a spiritual point of view, spiritual vision. If we do this we will of course have been thinking with the aid of our material organization, but we will have used this to reach a further stage of human development, ascending to the world of the spirit as a result. On the other hand we can stop at the point where as earth beings we let our mineral body do the thinking. It is perfectly able to do so. That in fact is the danger, and materialism cannot be said to be wrong in its views, particularly where thinking is concerned. This mineral body is no mere photographic print. It is able to think for itself, though its thinking is subject to the restraints of earthly life. We need to raise the experience our mineral body is able to give us into supersensible spheres.

It is therefore possible to say that it may indeed be true that human thoughts are merely something exuded by the human mineral organization. That may indeed be right, but human beings must do this in the right way.

Human beings have the freedom to develop on earth in such a way that they are merely the product of matter. Animals cannot do this; they do not get to the point where mineral incorporation leads to the development of thinking activity. Animals cannot choose to prove the truth of the materialistic point of view. Human beings are at liberty to prove the truth of the materialistic point of view; all it needs is the will to do so out of a materialistic attitude to life.

Human freedom is such that people are indeed free to make materialism come true for the human kingdom, that is, they can take a course that will lead to human beings on earth concerning themselves only with material things. Fundamentally speaking, therefore, it is a matter of choice if we become materialists. If people enact the materialistic attitude presented to humankind with enough insistence then they will make this outlook come true.

This influence on human beings comes from ahrimanic powers. They want to keep everything connected with earth evolution at the point now reached through the evolution of a mineral organization. They want to make human beings perfect, but only as far as their mineral organization is concerned. The luciferic powers want to keep human beings, who now have acquired a mineral organization, at the earlier stage that was right for them before they acquired a mineral organization. So we have two powers pulling at the traces, luciferic and ahrimanic powers. The luciferic spirits want to get human beings to a point where they finally cast off their mineralized bodies and go through an evolution that has no relevance in earth life and has merely been a stage within it. The luciferic spirits aim for the gradual elimination of everything relating to the earth from the whole evolution of mankind. The ahrimanic spirits aim to take firm hold of this earthly, mineral aspect of human beings, isolate it from progressive evolution and let it stand on its own. That is how luciferic and ahrimanic spirits are pulling in different directions.

It is absolutely vital that having presented the large outline we now come to apply this to ordinary everyday life. We do not consider a U-shaped bar of iron to be a horseshoe when it is in fact a magnet. In the same way we really should not consider human life to be entirely the way it may appear outwardly. If you shoe a horse with magnets you fail to realize that a magnet has more to it than a horseshoe. Yet it happens quite often nowadays that people speak of human life exactly like someone who shoes their horse with magnets rather than with horseshoes. People have no hesitation in speaking of positive and negative electricity in the inorganic sphere, or of positive and negative magnetism, yet they are embarrassed to speak of luciferic and ahrimanic elements in human life. These are just as active in human life as positive and negative magnetism are in the inorganic sphere. It is just that the idea of positive and negative magnetism is more easily understood. It does not take as much effort to grasp it as it does to grasp the idea that there are luciferic and ahrimanic elements. That is also the reason why we shall only learn to deal with the empty talk we hear today, empty talk that turns into lies, by knowing that it is luciferic by nature.

Similarly we shall only learn to deal with everything that shows itself here and there as the materialistic point of view by knowing that it is ahrimanic by nature. In future mere external characterization will not get us anywhere when we want to understand human life; all we would be doing is talking around the subject and committing the most foolish errors when we try to apply these external ideas to real life. One thing we would not be doing is deriving from our knowledge social impulses for social institutions established by human beings. This has very much to do with the utter seriousness required when looking at everything connected with evolutionary trends in humanity. We cannot gain understanding of the life we are now living unless we raise our vision from earthly concerns to spheres beyond this earth. Something singular is at work here.

Looking back into earlier stages of human evolution—though not as far back as those I spoke of earlier—people generally make judgements based on available historical documents. There are historians—well-known names—who say that the history of humankind consists of everything to be found in the written records. If you start from such a definition of history, like the historian Leopold von Ranke, you will obviously arrive at a particular kind of history. But writing itself is part of history, it evolved from something else, and in real terms one cannot do anything with this kind of definition.

We need only go back as far as Chaldean-Babylonian times, to Ancient Egyptian times, and we shall find that at that period of evolution human beings still related to the cosmos in a very different way. People today have no real idea of what it meant to connect one's life to the course of the stars, the planets, and their position relative to the fixed stars of the zodiac. These things have become an empty abstraction nowadays. Do you think a modern astrologer delving into ancient astrological writings—if at least he only searches through the old works, and does not produce new ones; the new ones are terrible!—has even the slightest idea of the living connection which the Ancient Egyptians and Chaldeans felt to exist between human beings and the movements and positions of the stars viewed from the earth?

Everything is different today. It has to be said that an important part of human evolution since those times has been the narrowing down of human awareness to the physical world. What did those Egyptians know of the earth? It was merely a portion of land. They knew more about the heavens. Their experience was one of upward verticality. The Ancient Greeks did not yet lapse into the horizontal plane either; they, too, gained their experience by ascending vertically.

The vertical became diminished as the horizontal started to spread. The maximum limitation human beings experienced in their knowledge of the heavens arrived with the great increase in knowledge of the earth that came when men sailed around the globe and found that having sailed away to the West they would return from the East. It was necessary for vertical human understanding to become obscured. Human beings had to be isolated from the universe so that they could find within themselves the only power that can lead to human freedom. Moral impulses will in turn arise out of this human freedom.

Human beings therefore no longer relate to the spheres beyond the earth in the vertical fashion of the Ancient Greeks and Chaldeans. We have had the schooling that only a horizontal plane can give and must now ascend again in moral, ethical terms. We must learn how human life is influenced by powers that do not show themselves in the unfolding of the world's outer existence. These are the luciferic and ahrimanic powers.

People tend to put their minds to other matters, however, and sometimes I must also tell you of things relating to our anthroposophically oriented spiritual movement. This has accepted the task of working out of the full seriousness of the needs of the time and listening to the language spoken from the cosmos beyond this earth, as it were, a language which tells us that we must once again come to see the way the human being is connected with the whole cosmos. But these efforts—and please forgive the abrupt change of subject—repeatedly encounter some very peculiar points of view taken by people who oppose our aims of furthering the progress of mankind. Let me read you a passage from a letter that is really typical. As I said, please forgive the abrupt change of subject but we are obliged to inform you of all kinds of things that are going on at the present time with the purpose of undermining and destroying this movement which endeavours to take up the challenge of the present age.

There is someone in Norway[3] who has made it his task to destroy our movement. To assure himself that he has a right to do so, this man is writing to leading figures—that is how one does these things nowadays. He wrote to a publication called *Political Anthropological Monthly*. This journal sent him the following information: 'Dr Steiner is a Jew through and through. He is connected with the Zionists, indeed associated with them, and works for the Entente.' The editor added that they—i.e. people of their kind—'have had their eye on him for some time'.

I just wanted to tell you this in conclusion, as yet another case among the many one encounters today, with new ones coming up almost daily. That is the attitude anthropologists are now taking toward the efforts being made in the domain of anthroposophy.

LECTURE 2

STUTTGART, 7 MARCH 1920

IT has been said on a number of occasions, and also two days ago when I presented the subject from a slightly different point of view, that it is important for us, in considering the evolution of humanity as spiritual-scientific research describes it to us, to grow aware of the gravity of the present moment, and act accordingly wherever we may stand in life.

Today I should like to add some further building stones to an edifice that, seen in its entirety, can show us the modern tenor of the human mind and spirit, and how we shall have to work for the further progress of humankind by taking this state of mind and spirit as our point of departure. To begin with let us refer to things which in the main are already known to us.

We know that the human race in its present state of civilization has by and large descended from the human race that evolved before disaster befell the continent of Atlantis. It has been said on a number of occasions that Atlantis occupied an area, between present-day Europe, Africa and America, that is today covered by the Atlantic Ocean. We know that under the influence of that disaster—in the course of its preparation and later as it proceeded—the peoples of that time migrated first in an eastward direction, populating Europe and then Asia as they moved on, and that the European and Asian peoples of the present day are in fact the descendants of the peoples of Atlantis. We also know that civilization then took the opposite route and people coming to colonize Europe brought with them cultural contents that had first been developed in Asia. These then spread further from a number of centres in Europe. Thus I would say that the physical basis for modern civilization is provided by the peoples of Europe and Asia, descendants of the ancient Atlantean race that moved from the West to the East. Civilization itself however moved from East to West. These two movements can only be properly distinguished through spiritual-scientific investigation. The two are confused in conventional anthropology and it is not realized that only the culture, the civilization, has been transplanted from Asia to Europe, whilst the physical basis comes from migrations that proceeded from West to East.

People always have some relationship to the locality where they live. We relate in some way to the soil under our feet, to everything this soil produces, to the way the soil comes to expression in climatic conditions and provides a habitat. You can conclude from this—and spiritual science fully confirms it—that the peoples who went further into Asia in the course of those post-Atlantean migrations inevitably developed in a different way from those who remained in Europe. In mundane terms this means that the soil of Europe had a different effect on the descendants of the Atlanteans than the soil of Asia. In a way, we can thus define the difference between the populations of Asia and of Europe. The difference is that, particularly during the earliest periods of post-Atlantean civilization, during the ninth, eighth, seventh and sixth millennium BC and the millennia that followed, the people of Asia adopted intellectual thinking, thinking as we know it, in a different way. This type of thinking did not fully emerge until the fifteenth century, as I said on the last occasion, but it was in preparation for centuries and indeed millennia before that.

This form of thinking as we know it today only developed in very recent times, assuming its true character as intellectual thinking in which the soul itself becomes inwardly active. But the whole of our evolution, particularly in post-Atlantean times, has been tending towards this intellectual approach. And it is significant that the post-Atlantean population of Asia took up all that we may call intellectual more into its soul elements. We can say that due to geographical conditions the peoples of Asia were specifically predestined to imbue their souls with the early stages of intellect. The most remarkable aspect of Asian civilization is that the soul element as such became the instrument for adopting the intellectual principle.

It was different with the people who had remained in Europe. Quite specifically the situation was that physical development, the physical organization, not only became the real instrument of intellectual development at a later stage, but evolved in such a way that at an earlier stage it already became the essential characteristic of these peoples, constituting itself as a particularly suitable vehicle for the intellectual principle. If we therefore wish to characterize the descendants of the Atlanteans' earliest descendants—ourselves in other words—we have to say that the Asian peoples accustomed themselves to thinking more with their souls; the Europeans accustomed themselves to thinking more with their bodies. That is in fact the major difference between the civilizations of Asia and Europe. If you want to show the clear difference between the kind of intelligence apparent in the Vedic writings or Vedantic philosophy and other cultural streams in Asia compared to European culture you have to say: Asians think more with their souls, Europeans more with their bodies.

The people of Asia may thus be said to have taken the intellectual element into a higher aspect of their human nature, with the result that an advanced civilization developed much earlier. This, however, was a civilization of the soul that had fewer abstract concepts, a culture that found its own ways to higher things, using the human soul and spirit to reach the soul and spirit of the world without resorting to abstract concepts. That is where the spiritual nature of Asian civilization lies—inasmuch as it is essentially a civilization based on soul qualities. The peoples of Asia largely left their bodies unused when it came to thinking; they merely carried their bodies with them through life on earth. The life of the mind was nurtured entirely at soul level. You cannot understand the singular nature of Asian culture unless you look at it from this point of view.

Europeans based their thinking increasingly on the physical body. That is also why the foundations were more strongly laid among them than in Asia for a culture in which freedom can be the central principle. The people of Asia, endowed with intelligence at soul level, remained a more integral part of the whole cosmic organism. The human body isolates itself especially from the rest of the cosmic organism. Using it as the instrument for our intellectual life we become more independent, though this independence is more bound up with the body than is the case with the people of Asia who have developed intelligence within the soul principle and are consequently less independent.

As the time approached in the history of humankind that was to bring the Mystery of Golgotha, an advanced culture of soul and spirit had evolved in Asia. At the time of the Mystery of Golgotha it had already reached its culmination and was in the early stages of decline. Do not let us deceive ourselves: European ideas do not make it easy to grasp the great culture which had grown out of the soul and spirit of the Asian peoples. When people like Deussen[4] who are thoroughly European in their way of thinking, people for whom the physical body is the instrument of thinking, try to communicate the nature of Asiatic culture to Europeans, such efforts fail to embody the content of Asian soul-spiritual civilization, for everything alive in it has been translated into European thought. Such interpretations of certain Indian spiritual streams have sometimes even caused a sensation in Europe—those published by von Garbe[5] for instance—yet this was nothing more than European materialism producing a garbled translation of the nature of Asian soul and spirit. Publications of this kind contain no trace of the real spirit of Ancient Asia. It is necessary to point this out very clearly because, as I have said before, people nowadays have an extreme belief in authority and really only acknowledge the validity of something if it has been written by university professors. There is of course no real inner reason to regard Deussen's or Garbe's botchwork as of any importance except as an indication of this erroneous belief in authority in Europe. People are no longer able even to find inner reasons; they merely believe one thing or another to be right because some outer authority says so. It is not worth keeping silent about such matters— even if it means making more enemies—for the gravity of the present situation absolutely demands that there shall be no compromise where certain things are concerned, and that the truth must be clearly stated.

The advanced culture of the spirit in Asia was already to some extent in decline when the event of Golgotha occurred. This event of Golgotha—it cannot be sufficiently stressed—was first of all absorbed and understood by minds that were the product of Asian culture. It is important to distinguish between the Mystery of Golgotha as an historical event that happened in the Near East at the beginning of the Christian era and the notions people have of this Mystery of Golgotha. At the time when the event occurred, Europe did not have the capacity to grasp it fully, for it was an event that could only be grasped in soul and spirit. European civilization, however, had spread by using physical matter as its instrument. The event which occurred at the beginning of our Christian era could not be directly grasped in a civilization based on physical and material things. Asian civilization on the other hand had an intellectual life based on soul and spirit and out of this was able to find concepts with which to grasp the event of Golgotha. The event that happened in Palestine was thus poured into the conceptual world of the Orient. Thus it travelled westward through Greece and Italy and was brought to Europe in the form of tradition.

People can be given something in an external way that they cannot yet grasp in their hearts and minds. Things may come to them in the form of a tradition or through the written word. Europe initially received its explanation, its understanding, of the Mystery of Golgotha, from the Oriental tradition. Christianity was understood in the light of Oriental wisdom, a wisdom of soul and spirit that was truly great at a time when the Mystery of Golgotha, and the way Christ related to the whole of earth evolution, were still perceived gnostically. This wisdom dwindled more and more as Europeans informed this tradition with their own unique characteristics. The way they saw these things inevitably came to be informed by the specific characteristic of an intellectual life bound to the physical body. In early times the human bodies of Europeans were very much the instrument of their kind of elementary intellectual thinking, but then this body gradually began to die away. Until the fifteenth century, and even right up to the present, the physical evolution of European humanity involved the physical body becoming increasingly deadened. Our physical bodies are growing ever denser, ever more ossified. We cannot demonstrate this through ordinary anatomy and physiology, but it is true. We no longer have bodies as inwardly alive as those of people living in the first, third and even the tenth and eleventh centuries. Our European bodies of today have grown ossified, deadened, compared to those ancient bodies that were inwardly alive. Thus you have on the one hand a tradition apt for the soul and spirit of Asian people who preserved the ecclesiastical creeds, and on the other hand a European body for which those Asian traditions were increasingly alien, and which in the end no longer found itself able to receive the ideas coming from Asia.

From the middle of the fifteenth century onwards the influence of the ossified European body has been such that in the end that ancient tradition only survived amongst religious denominations in empty outer phrases. For many centuries the tradition remained so alive that little regard was paid to the Gospels and people took their cue from life itself. As the European body grew ever more less alive, people felt impelled to cast off the old traditions and put their faith only in the word, the written word. People believe they have the Word when in fact they only have a poor translation of it. It gradually came about—though no one is willing to admit it—that really all they had was the outer shell of the Word of old that once held within it the tidings of the Mystery of Golgotha in the garb of Oriental wisdom, a wisdom of soul and spirit.

This Oriental wisdom is little understood by the people who generally interpret or translate the Gospels; they understand little, if any of it. The point is that it is necessary to see the Mystery of Golgotha in a new light. However, unless we Europeans get beyond what a dying physical body is able to give, we will be unable to do so. We must develop through spiritual science and come to grasp the spiritual world in a way that is independent of this body. Our future salvation entirely depends on our ability to grasp the spiritual world in this way, independently of our physical bodies, connecting directly with the spirit. This insight will have to be different in essence from the Oriental culture of soul and spirit, which came as though of its own accord through human evolution. Europeans of the present time will need to achieve it by their own efforts. They will have to cultivate spiritual science. They will have to create an educational system in which, from kindergarten onward, spiritual science is not presented as a theory but flows into everything we do as we teach and educate children. Spiritual science should also flow into higher education and should be alive in everything connected with art, literature and so on, everything that is our common cultural life. This European culture must provide for the nurturing of spiritual science itself. On the basis of such a spiritual science the Mystery of Golgotha will then be seen in a new light. We shall have to recognize that oriental soul-spiritual interpretation of the Mystery of Golgotha belongs to ancient times A new wisdom will have to grasp the Mystery of Golgotha in a new and living way.

We have often spoken about these things. However, they need to imbue our souls in every way. We must come to feel deeply what a serious matter it is to realize the need for new insight into the Mystery of Golgotha. And in Central Europe, in particular, there is something that renders this a still more serious matter.

Looking with more profound insight at what has become our culture in the second half of the eighteenth and first half of the nineteenth century particularly in Central Europe, you really have to say this: Though the bodies of people in Central Europe were already growing more deadened, there was still sufficient living quality present to enable people to rise to a world of ideas more alive than ever seen before in the evolution of humankind. Nowhere else did human minds rise to abstract ideas in such a way that, within these abstract ideas, they were not dwelling in the sphere of death but in the sphere of life. This was achieved in Goetheanism, for instance, and by the philosophers of German Idealism. It is not something to be found anywhere else in human evolution. It was, as we need to fully recognize, a high point, a culmination. People today no longer want to accept that Schelling, for instance, to take just one of the Goetheanists, moved in a sphere of abstract thoughts and yet, whilst speaking in quite abstract terms, moved in the sphere of abstract thoughts in a living way: in a way as alive as people usually are when they speak of food and drink. The same is true of Fichte. In this realm of human evolution we can become especially aware of an ability to ascend into the sphere of concepts and ideas in a very living way. Something quite special exists therefore for this Central European population, special in the whole context of human evolution in more recent times. They have their own characteristics that enable them in particular to take up the vocation of humankind for the present day: namely, to enter into spirituality again. Only in the second half of the nineteenth century did these characteristics come to be drowned out again by other things.

It is terribly painful to become aware of such somnolence in human nature in Central Europe today, sleepwalking in a sense over the graves of Lessing, Goethe, Herder and Schelling. This human nature considers such soul sleep as intrinsic to its task. If we were to develop the thread of the writings and thoughts of those great minds, not externally but by entering into the spirit in which they wrote, we would find the element that can raise Europe to the heights. Europe cannot be raised to the heights by Gospel words that no one understands repeated parrot-fashion in the churches. European culture can only reach its heights if people seek to grasp the spiritual worlds by elaborating further what Herder, Goethe and others were working towards. There is, however, scarcely any awareness of this at the present time. It is a sad sign of the times, for example, that in a culture which possesses treasures such as Fichte's *Destination of Man*, Schelling's *Bruno* and Schiller's *Letters on Aesthetic Education*[6]—as well as many more I could mention—we end up with the inane and superficial Americanisms of Ralph Waldo Trine[7] and the like. We have things that are much, more sublime but we let them sleep and turn to other impulses.

The further we penetrate into the actual life of the mind and spirit the more it becomes apparent that something new is emerging in the life of humankind today. Central Europeans are far from understood by Western Europeans; Western Europeans are far from understood by Central Europeans, even in ordinary life. People are not aware of this, however. They think they understand each other. They do not realize that they are not communicating their thoughts. I am not referring to the way Americans and Europeans fail to communicate, but Central and Western Europeans. You come across some odd things in this respect.

The last time I was here I told you that vilification is rife not only within Germany but also beyond its borders.[8] I told you about this man Ferrière,[9] for instance, who spread one of the strangest tales in a Swiss-Belgian journal, saying that it was of course generally known that I was 'Rasputin' to William II[10] and had a major share in all the bad advice William II was given back in those terrible days. This slanderous story came to be widely known particularly in French-speaking Switzerland and I therefore defended myself by writing down the truth of the matter, stating the bare facts. I said that I had only ever seen the former emperor briefly and from a distance, had never spoken with him at all and never sought to contact him, even through others. These are the bare facts I stated in a letter to Dr Boos[11] who then passed this on to Mr Ferrière. The matter was published in the journal in question together with Ferrière's reply. This went more or less as follows: once again the great difference between the Latin and the Germanic mind is demonstrated. The Germanic mind takes everything so seriously. 'My readers,' Ferrière wrote, 'will of course not have been deceived; they will have realized that what I wrote was intended light-heartedly, as *plaisanterie* rather than *méchanceté*.[12] Apart from that, let me state that one can learn that something we may have heard from people whom we think we can believe need nevertheless not be true, even if it is a widely believed rumour. I take note of this.' And so on.

That was the elegant reply the 'Latin mind' was able to give, a 'pleasantry' concerning the Germanic mind. It is a good thing at least when such things come to the fore; very often they are not even noticed.

They assume even greater significance where a more profound view is taken of the world, at the point where they relate to initiation knowledge and everything connected with this. This is a sphere where we do, for once, have to mention these things, though some people consider it highly dangerous to touch on them even today. I want to talk to you today about a matter that in the opinion of representatives of initiation knowledge, Western representatives in particular, ought not to be discussed.

Western representatives of initiation knowledge will tell you again and again that it simply will not do for anyone to disseminate initiation knowledge which they themselves have gained. You will find that when genuine initiates in the West present initiation knowledge in books available to the public, they always deny having personal experience of the things of which they are writing. You will find that it is quite typical (such books have appeared particularly in America) to have a Preface—this is part of the whole technique—which states that 'None of these things are my own, of course, for if they were my own I would not communicate them.' Take a look; you will find this kind of thing in many testimonies published particularly by Western initiates. If you ask why this is done you will be given an answer that, within certain limits, does hold true for Western initiation science. You will be told that anyone learning something directly from the spiritual world, who knows the secrets of the spiritual world, cannot tell another person that they have gained this through personal experience—such words will often be used—for if they betray the fact that they personally possess initiation knowledge they become dependent for life on the person to whom they have betrayed their secret.

The effect of this attitude, which has its roots in the intrinsic nature of Western initiation, is that everything to do with initiation is discussed in a very superficial way in Western initiation fraternities, and that there are indeed unknown initiates amongst Western humanity. This is an attitude that has to be overcome in the new age; it cannot hold true in Central Europe, and the spirit which must arise there will have to combat this attitude. It will have to do so by coming to understand the Mystery of Golgotha in the new and spiritual way I have talked about. It will have to come to understand the presence of the Christ in human life.

Here lies a major secret. Customary initiation knowledge in Western countries is far removed from Christianity; otherwise the Theosophical Society would not have excluded or caricatured the Christian faith and dressed up a purely oriental, pre-Christian Indian wisdom as something new. It is a peculiar characteristic of this Western initiation knowledge that its initiates only gain something from their initiation if they have at least one pupil who reiterates their ideas. There is no point whatsoever in having initiation knowledge just for oneself. If your eyes look straight ahead you will not see a single object. In the same way you will not encounter your own spiritual ideas as a Western initiate unless you can see them repeated by someone else. There are many indications of this, but it is not properly realized.

This being so, it is indeed true that someone who betrays to another person the fact that they are an initiate will be in the other person's power for the rest of their life. This is because the other could then refuse to serve them, refusing to repeat their ideas. This principle implies some degree of dependency, which is by and large a characteristic of the initiation knowledge I have frequently referred to in other respects as the prevalent form of initiation science in the West.

There is only one way out of this dependency on one's followers and that is to be in communion with Christ, who can truly be found on earth since the Mystery of Golgotha. We will then be in communion not with a human being who happens to be imperceptible to the senses, but with the first among brothers who has come among humankind, the living Christ walking among us. If we are in communion with Christ as we had to be in communion with other human beings in pre-Christian initiation, we need not be afraid to share our own wisdom with our fellow human beings. Today there is no other way in which original initiation wisdom can be directly communicated than by being in communion with Christ. There is no other way. A genuine initiation wisdom of the present age will have to look for such communion with Christ.

If this initiation wisdom were not there to be found we could make no progress in social understanding. It is no longer possible to develop social ideas nowadays unless we make initiation our foundation. Yet we have need of social ideas. A social system born out of Western initiation wisdom alone would depend on that initiation wisdom being kept secret even at its lowest levels—certain higher levels cannot be made known today because people must have the necessary preparation. Keeping things secret in this way is not compatible however with the principle of people being equal, a principle modern Europeans and indeed the whole civilized world consider important today. So you see that precisely when it comes to initiation wisdom, a colossal difference becomes apparent between the Central European and the Western mind. The difference becomes even more apparent in the case of initiation wisdom where people talk at cross purposes with each other as they do today, and believe that humankind can be brought into abstract uniformity. That cannot be done. Human beings are differentiated and this differentiation shows itself particularly if one takes a more profound look at initiation knowledge. This is an important subject and it will no doubt be necessary for me to explain it in more detail while I am here. When it comes to genuine spiritual insight one simply cannot be slipshod about things, and a lack of seriousness concerning the truth is unacceptable. It simply will not do. Truthfulness is of the essence.

Lecture 3

STUTTGART, 9 MARCH 1920

THERE are a few things I want to add to the points we have been considering. They may help to make some of the ideas on which we must base our actions more real. I am looking for ideas that are less abstract than the vast majority of ideas by which people allow themselves to be governed today. We really need such tangible, concrete ideas, for they are the only ones that enter into the realm of human feeling, and therefore into real life. They are the only ideas to fire the human will and human actions.

Looking at the world today we should really consider the most striking characteristic of social life in the civilized world to be that the smaller communities of past times have given way to quite large human communities. We need not go far back in human evolution to find that social communities encompassed only limited territories. Towns and cities were, relatively speaking, a whole and single community, and really it is only now, in quite recent times, that large empires have arisen, that the empire of English-speaking peoples has come about—I have characterized this a number of times. None of us should have any illusions about the consequences of this, particularly in Central Europe. Yet to form the right ideas about these things, ideas that fully relate to reality, we need to adopt spiritual-scientific perspectives. The spiritual-scientific point of view takes us back to earlier stages of human development to see that then, too, people formed certain kinds of communities, though these should not be called 'states'—as I have said on a number of occasions—for that would cause tremendous confusion. Instead, let us find some other, more neutral term. Let us say that 'realms' arose. Such realms were ruled by individuals, or by particular groups. Subsequently nation states developed out of this, and today these nation states are taken so much for granted that no one would think of questioning them—at least, certain aspects of them are not questioned. What is worse, they are so much taken for granted that people are not even inclined to think about them.

Behind this, however, lies something that unites human beings' inner soul life with the spiritual, the divine principle, as it came to be called during the different stages of earthly evolution.

If we go back to prehistoric times, times that only partly extend into historical times, we find that in those prehistoric eras the idea of a ruler of the realm, as we may call it—whatever words we use do not really fit those earlier ideas—was quite different from what we take it to mean today. The idea of the ruler of an earthly realm came very close to people's idea of a god. These things inevitably seem highly paradoxical to modern minds, though that is only because such minds are disinclined to give serious consideration to things that existed during past eras of human evolution and do not accord with the way of thinking that has become customary in Western Europe and in its appendage, America, over the last three or four hundred years.

Of course the way a ruler of the realm was introduced to his office, at least in many empires, was very different in those early, partly prehistoric times. We need only go back as far as Ancient Egypt, meaning the earlier, partly prehistoric times of Ancient Egypt, or to Ancient Chaldea, and we shall find that it was considered a matter of course for regents to be prepared for their office by the forerunners of our present-day priests. People had quite specific ideas as to how a ruler should be prepared for office by the priesthood and its institutions. They felt that with this kind of preparation the person called to be regent truly became what the Chinese, still having a faint notion of this, called the Son of Heaven. There was an awareness that someone called to rule over some region had to be made a kind of Son of Heaven. What was in people's minds was something quite different, however, from the idea we have today when we speak of training a person or preparing them for something. You can go to great lengths to explain that the whole person needs to be developed, that it is not adequate to train people for some office or other by merely inculcating them with intellectual knowledge; practically all our ideas today on development, education and so on tend to be abstract to an extreme degree. People have the idea that only a particular aspect of the human being should be changed or transformed to train them, to prepare them for some office. No one considers that development should bring about a complete change in an individual. Above all no one thinks that something objective should enter into the soul of that individual, something that was not there before. No one conceives of this. I could characterize this more or less as follows: the person I talk to is a product of the natural and social conditions of the present day. He tells me this and that, I tell him this or that. The person who speaks to me bears a name: he is the product of the usual natural and social context that prevails today. The same applies to me. That is really almost the only way in which we behave towards each other nowadays, the way in which we look at each other.

In the times of which I have been speaking that would have been a very alien notion. It was above all alien to those called to hold important offices, those invested as leaders within the social community. Their outward natural background—family origins, father, mother, grandfather, grandmother and so on—was of no real concern if the people concerned had been properly prepared for their office. The things we look for and find in modern individuals, including those summoned to the highest offices, were then of no account. People felt that if they spoke to someone who had been properly trained in this respect it was not the ordinary ego that spoke to them: i.e. an ego born in some place or another, bearing the imprint of some social background or other. Instead they felt that something was speaking to them that had been brought down from spiritual heights to take up its abode in a human individual thanks to the preparation and training given in the mystery rites. This must of course sound incredibly strange to people today. But we need to stop harbouring confused notions about these things and develop ideas founded in truth.

The idea was that education—not all training but the training of people called to high office—should enable spirits from the higher hierarchies to speak through these individuals, using them as their instruments. Those instruments were prepared by schooling, so that spirits of the higher hierarchies would be able to speak through them. There was general awareness of this, particularly when the population at large thought about their ruler. Remnants of this still survive, for instance in the title Son of Heaven for the ruler of China. That was the nature of human consciousness as spiritual science finds it to be in earliest Egyptian and Chaldean times. To the people at large their ruler was a god. Basically they had no other concept of divinity. The preparation of the ruler was such that his outer physical form was of no significance, merely the vehicle for a god to move among human beings. The earliest inhabitants of what was later to become the kingdom of Egypt quite naturally accepted the fact that they were ruled by gods who walked on earth in human form. In this respect the earliest social awareness of human beings was, in one sense, entirely realistic. There was no recognition of a separate world beyond, of a separate spiritual world. The spiritual world existed in the same place as the world within which people moved on earth. In this world, where human beings walked the earth, not only ordinary human individuals were walking about in physical form but also gods. The divine world was there amongst them, made absolutely visible under the conditions they were accustomed to creating through the mystery rites. When such a ruler wanted something, decreed something, it was a god who wanted it. To the minds of earliest humankind during that semi-prehistoric period it would have been pointless to question whether something decreed by their ruler should or should not happen; it was after all a 'god' who wanted it.

Earliest humanity thus connected the spiritual hierarchies with everything that happened on earth. Those hierarchies were there in their midst; they were not something to which one first has to ascend by some kind of spiritual, inner means. No, they were present in the mysteries as the schooling given to physical bodies found suitable for preparation as dwelling places for spirits of the higher hierarchies, so that these might walk among human beings and be their rulers.

This may seem strange to modern minds, but it is high time that modern minds left behind the narrow-minded views they hold today—ideas, as we know them today, only three or four hundred years old—and take a wider view. We cannot develop and think ahead to the future unless we broaden the tunnel vision which has evolved in almost every sphere of life. We must expand the time horizons we survey and consider larger evolutionary time spans than modern history normally covers.

The things of the past, things that existed in historical and prehistoric evolution, do of course give way to other things as time progresses, but in certain areas they are retained. They are often retained by becoming external, continuing in an outer form and losing their inner meaning. The awareness of the godlike nature of the ruler that was a feature of the earliest empires still figures here or there in the present age, except that it no longer has any meaning, since mankind progresses and does not stand still.

Not long ago a Roman Catholic bishop addressed a pastoral letter to his diocese[13] in which he stated nothing more nor less than that a Roman Catholic priest conducting an act of worship was more powerful than Christ Jesus himself. Acting as the celebrant, the priest compelled Christ Jesus, the god of Christianity, to enter into the physical world as the priest performed the act of transubstantiation. Whether willing or not, the act of transubstantiation forced the god to follow the route prescribed by the priest. This pastoral letter did indeed refer to the sublime power of the earth-born 'priest god' over the 'inferior god' who descended from cosmic heights and walked on the earth in the flesh of Jesus. Things like this have their origin in older times and have lost their meaning in the present age. Some people representing certain confessions know very well, of course, why they keep introducing such things into human minds. They have become as meaningless as words a modern ruler inscribed in an album, stating that 'The king's wishes are the supreme law.'[14] We have witnessed these things in our times. Humanity, fast asleep, has said nothing at all about them; and people still say nothing about things that bode ill for humankind, things they simply accustom themselves to, that they refuse to examine—in the same way that they refuse to recognize major milestones in human evolution.

That was a first stage in the evolution of human empires: when the ruler was a god. This outlook was still very much alive in Roman times. Whichever way you look at Nero, as a fool or a bloodhound, for the large majority of the Roman people Nero's dreadful tyranny merely made them marvel that a god could walk on earth in such a guise. Many of the citizens of Imperial Rome never doubted that the figure of Nero was that of a god.

A second stage in the evolution of empires came with the transition from a ruler who was a god to one who ruled by the grace of God. During the earliest times of human civilization, the ruler was God. At the second stage the ruler stood *for* God; God did not indwell him but inspired him, endowed him with special grace. Everything he did would succeed because divine power—now no longer within him but in a realm close to the earthly realm—flowed into him, inspired him, filled him, and governed all he did.

To describe the essential nature of such a ruler of this second stage we have to say that he is a symbol. At the first stage the ruler was a divine spirit walking on earth. At the second stage he is what that spirit signifies; he is the sign, the symbol in which the spirit comes to expression. The ruler becomes the image or representative of God.

The principles governing those external social relationships also came to expression in institutions. In earliest times empires were so constituted that a number of people were governed by a divine spirit. This god would be similar to them in external appearance but utterly different within. At the second stage we find empires where the leader or leaders signify the god or gods, are symbols of them.

At the early stage of human empires it was beside the point to question whether the ruler, the god, was acting rightly or wrongly. At the second stage it began to be possible to think whether something he had done was right or wrong. At the early stage everything the ruler did, thought or said was right, for he was the god. Then, at the second stage, it was felt that some other spiritual sphere exists side by side with the earthly realm that contains within it the god, the ruler possessing the grace of God. The power streaming into the earthly realm, giving direction and orientation, was felt to come from that other realm. The institutions and human individuals in the earthly realm were the reflection of something streaming in from the realm of the higher hierarchies.

It is interesting to find, for example, that Dionysius the Areopagite,[15] also called the pseudo-Dionysius, who was much more authentic than orthodox scholarship imagines, proposes the right theory about how human empires are ruled by divine empires so that the conditions and institutions created among humans are a symbol of what exists in the divine realm. Dionysius the Areopagite wrote that there were heavenly hierarchies behind the human hierarchies on this earth. He stated very clearly that the social structure of the priestly hierarchy here on earth, from deacons and archdeacons all the way up to bishops, ought to show that the relationship of deacon to archdeacon is the same as that of angel to archangel, and so on. The earthly hierarchy should truly reflect the heavenly hierarchy. This relates to the second stage of empires. Something evolved that was to govern human ideas until quite recent times. After all there existed in Central Europe until 1806 an institution that in its title gave expression to the way the heavenly and the earthly principle were seen to be one: the Holy Roman Empire, an empire seen as being based on the power of heaven. To this were added the words 'of German nationality' to show that the empire was also of earthly origin. The way the title evolved makes clear that a whole empire was formed in a way intended to be seen as the image of a heavenly institution.

Such were also the ideas behind St Augustine's *City of God* and Dante's work *On the Monarchy*.[16] If people's ideas were less narrow they could take a broader view of something like Dante's work and realize that Dante, who after all must be considered a great thinker, still had ideas in the thirteenth and fourteenth centuries that are radically different from our modern ideas. If we were to take such things in historical evolution seriously we would give up those limited ideas that do not even go back as far as Dante but are just a few hundred years old. The illusory ideas with which people nowadays fill their heads, wanting to understand history merely by going back as far as Ancient Greece, are limited ideas. Yet it is only possible to understand the whole structure of Ancient Egypt, for instance, if we know that the ancient gods still walked on earth then. In the times that followed, gods no longer walked on the earth, but the institutions created on earth had to be an image, a symbol, of the divine world order.

Then it became possible to reflect, for instance, on the lawfulness of things, to reflect on such things as the fact that the human intellect can arrive at a judgement as to what is lawful and what is not—all this only became possible during the second stage of imperial development. During the earliest stage it was pointless to reflect on what was lawful and what was not. People had to observe what the ruler dictated, for the god lived in him, he was the god. In the second stage, human judgement could be used to determine that there is something in a spiritual realm adjoining our own realm that we cannot reach as physical human beings but only as human beings of soul and spirit. Then people no longer believed, as they had believed in earlier times, that the divine could unite with the whole physical human being, that a human being could indeed become a god. At most—using mystical language to define what lived in public institutions—people believed that the human soul could unite with the god.

Basically it is true to say that no one nowadays is able to understand the way things were expressed in works written and published as late as the thirteenth and fourteenth centuries unless one knows that the people of that time had quite a different awareness, a feeling that some degree of divine inspiration was alive in those who were called and trained to hold special office. Oddly enough, terms for serious matters will often become derisory expressions at a later point in human evolution. Someone nowadays using the expression a 'god-given right' would say this more or less ironically, or as a joke. Yet though it may be said today partly as a joke, in times when empires were at their second stage of development it certainly held true and was at the forefront of people's minds. It applied not only to people but within certain limits also to what was being done.

The actions performed in rituals reflected what occurred in the spiritual realms. The rituals performed were spiritual events reaching down into what happened in the physical world. People certainly believed not only that the spiritual realm adjoined the physical realm, but that it also worked into the earthly realm, and that the symbol or sign of the spiritual realm was to be found in the earthly realm.

Very gradually people ceased to believe in the truth of this. An age was to follow where this awareness of the connection between earthly and spiritual things would fade. In the days of Wycliffe, and Huss,[17] people began to dispute things which it would have been unthinkable to dispute before. They disagreed about the significance of transubstantiation: whether this ritual act had anything to do with what went on in spiritual worlds. When people begin to disagree about such things, old ideas are coming to an end. People no longer know what to think; yet for centuries, indeed millennia, they had known exactly what to think. It always happens that certain things normal to a particular age continue to play a role in later ages. They are then out of place, anachronistic, luciferic. That is what happened to the great, far-reaching symbols of a particular age that testified to the connection between ritual acts and suchlike performed on earth, and divine, spiritual occurrences. Those symbols persisted during later ages and certain secret societies preserved them in a luciferic way. Western secret societies in particular preserved such ancient symbols. They are traditional in those societies, though they have lost their real content. On the one hand, then, we see certain secret societies that gave rise to the Freemasons, Jesuit organizations and denominational groups preserving, one can say, the symbols which only had meaning in a former age. But we also see how words preserve ideas that belonged to an earlier age in a basically luciferic way. Words used in everyday life have also lost their original substance, lost the context of a way of thinking when those words were symbols of spiritual things. Spiritual content is gradually lost and words become empty symbols, signs without meaning.

During a third period, at a third stage of empire, there was no longer any awareness of individuals receiving the grace of God, of divine elements entering into earthly events, earthly speech.

The spiritual realm now shifts entirely into another world, a 'beyond'. The opposite of what had existed at the first stage of empire now holds true. During the first stage the god lived on earth, went about in human form. During the third stage one can only think of the god being present in an invisible world that is not perceptible to the senses. All the expressions people used for their relationship with the realm of divine spirit lost its meaning. The word 'god' continues to be used. When the word 'god' was spoken in the distant past people were looking for something that appeared in human form, walking among human beings. It is not that human beings in those very early times were materialists. Materialists only became possible once the spiritual realm had been banished from the sense-perceptible world. During the earliest period of human evolution the spiritual world was immediately present among human beings. There would have been no need for someone living in Ancient Egypt during its earliest times to say, 'The kingdom of heaven is at hand'. That would be taken as self-evident. At the time when Christ Jesus appeared amongst humanity, people first had to be told: 'The Kingdom of the gods cometh not with observation... for, behold, it is within you.'[18]

We are now living in an age when it has become nonsensical to see anything in someone except the linear development of what they were as a child, a development based on cause and effect. We live in an age when it would be sheer madness for someone to consider themselves more than the straightforward further development of what existed in them during childhood. Something taken as a matter of course eight thousand years ago, as the universal outlook, becomes madness to utter nowadays. The realities of those distant times have been reinterpreted in the modern way of thinking in that convenient fiction we call 'history'. This spreads a veil over the radical metamorphosis we can discover when we examine human evolution in a way that accords with the truth. The things we often say today, things we reveal in our external life, exist because they once had relevance and were considered to be the truth. We still say things like 'by the grace of God'—people have more or less tried to get out of the habit in recent years, but they have largely failed to do so—but we do not know, or pay no attention to the fact, that there was a time when this was a reality for human minds, when it was taken as a matter of course.

I am pointing to things that give our public life its meaningless, conventional character. Things expressed in the words we use, in our customs and indeed the way we judge issues in public life, relate to times when those words, even if they only became part of the vocabulary at a later date—reflecting their linguistic origins—were formed and used quite differently. The words we use in public life today have been squeezed dry. With some it is immediately apparent, with others it was not apparent for a long time. In the distant past a token was hung upon a human body in magical rites, transforming it into an important magical aspect of the god walking on earth. This has become something trivial in the decorations given to people today. This is something people largely fail to recognize. Not only words can become empty phrases and lose all meaning, as is the case with the most important words used in public life today; the objects hung around people's necks or pinned to their chests can show a similar character in their relation to reality, like a word that is meaningless today but once had sacred meaning and substance.

We must realize that initially our evolution was such that an older awareness lost its substance and became empty and conventional. There can be no real new growth in the ruins of public life today unless we do realize it. We must look for new wellsprings that will give real substance again to public life. We have no awareness of gods walking the earth in human form. We must therefore acquire the ability to recognize the spirit not in physical human form but in the form it has when we rise to spiritual vision. For us the gods no longer descend to sit on physical thrones. We must acquire the spiritual faculties that enable us to ascend in our vision to the thrones where gods are to be found who can only be alive to us in the spirit. We must learn to fill the abstract formulas we use with spiritual contents of our own experience. We must become able to face truths that are deeply disturbing to those who grasp them rightly. We must become able to see things as they are. Sometimes we fail to do so for many decades.

As Central Europeans we believe ourselves to be part of European civilization. We should ask ourselves what it is that has made our inner life, the life of our soul, so full of discord over the last fifty years or more. Let me say just one thing. When you look westwards you see in the first instance—we'll leave the rest aside—a nation falling into decadence, the French nation. One thing is important, however, among the French. When a member of this nation said: I am a Frenchman—as they have said for centuries—he said something that accorded with external realities; a permissible, truthful declaration made with reference to outward life. Any of us who have ever talked to Germans who were young in the first half of the nineteenth century will be able to confirm the following. Herman Grimm[19] for instance has repeatedly described what it meant, to people who were young when he was young in Germany, for someone to profess himself to be German in public life, not as an empty phrase but in reality. It would have been treason. People were Bavarians, Wurttembergians, Prussians, but to say 'I am German' would make them criminals. In the West there was a point in saying 'I am a Frenchman'; it was permissible to be this in outward life. But to say one was German back then, that in other words one belonged to a united nation, would have meant going to prison or being beyond the pale in some other way. People have forgotten this now, but those are the facts. And it is important to face up to these things. We shall not develop the right enthusiasm for them, however, unless we enrich our inner life by considering the great events of world history, and seeing them in the right light—not that fabricated tale written in our textbooks and taught in our schools today, but the true history of the world that can be found by looking at things in the light of the spirit.

It is quite unthinkable for a modern Protestant Christian to acknowledge how people once found meaning in the thought that 'God walked on earth in the form of the ruler'; and then also that 'There is no kingdom on this earth where gods are still to be found, for the ways in which a person becomes a god now live in the supersensible realm, within the Mystery'. In the early times of Egyptian history, which in part was still prehistoric, the Mystery was still a supersensible reality. It was only when the mysteries were made into Churches that the Church became symbolic of the supersensible.

Modern humanity has no wish to look back to its historical beginnings. It lives like someone who has reached the age of forty-five and has forgotten what life was like as a boy or a girl, at most remembering back to their twenty-fifth year. Try to visualize what it would mean for the inner soul life of someone of forty-five to remember nothing that happened before the age of twenty-five. Yet that is the state of mind humankind is in now, and from it emerge those who wish to be the leaders of humanity. In this state of mind people attempt to find some direction and orientation that will inform a social system. But the most important thing is for people to see humanity as a living organism, one possessing a memory that must not be ridden roughshod over, a memory reaching back to things that still take effect to this day, but in such a way that they urgently require us to imbue them with a new quality.

Allowing this thought to resound in us a little, we shall see that there is need for something today compared to which all the empty words and phrasemongering are worthless. It would be good if a sufficiently large number of people were to realize how serious the present situation is, and out of this realization were to arrive at something that is really new. The sad thing is that people today face great tasks and yet would prefer to sleep through them. The particular task given to the anthroposophic movement for decades now has basically been to shake humanity out of its sleep, to point out that humanity needs to receive something today that alters the present condition of soul in the same way as we are woken from sleep and dream in the morning, and transform this dreamlife into the full wakefulness of day.

This is how I wished to conclude these two lectures on history in the light of spiritual science. It is my fervent wish that something might emerge from our anthroposophic movement that can truly fire our social ideas, filling them with warmth and energy. Social impulses are needed today, as is obvious when we look at events that we really should not overlook. These social impulses can only come to fruition if a new spirit is poured into human evolution. People who join the anthroposophic movement, whatever their background, need to recognize this. On the foundations of this anthroposophic movement truthfulness and alertness are necessary, real alertness.

The public life of modern civilized society has fallen asleep. Today people are so deeply asleep, and pursue their affairs in such external ways, that—were it not for the fact that one stands within spiritual life and perceives the course taken by spiritual affairs behind the physical world—it would be very easy to become perturbed. The external course people pursue clearly shows that they avoid looking for the truth in all that occurs. They are pleased not to have to scrutinize what is really happening. You can see that when people are told of something happening somewhere today, they stand there unperturbed and give no indication of having heard something of profound significance for the way things are going. People hear of deeply significant things that must inevitably lead to ruin, to decline and fall, and they do not even feel indignation. Things are going on in the world, intentions are alive in German lands, that should horrify people—yet they do not. Anyone incapable of being horrified at these things also lacks the power to develop a sense of truth.

It must be said that healthy indignation about things that are unhealthy should be the wellspring of enthusiasm, of the new truths that are needed. It is actually less important to convey truths to people than it is to bring fiery energy into their lethargic nervous systems. Fiery energy is needed today, not mystical sleep. Longing for mystical peace and quiet is not of the essence today but rather dedication to the spirit. Union with the divine has to be sought actively today, not in mystical indolence and comfort.

This has to be pointed out. We must find a way of making it possible again for our minds to connect a divine principle with our outer reality. We shall only be able to do this if we consider without bias how people found their gods walking the earth in the kingdoms of the earliest times. We must discover the way to enable our human souls to walk in the spirit in spiritual worlds, so that we can find the gods again.

Lecture 4

STUTTGART, 13 JUNE 1920

ONE particular fact that we have discussed here a number of times is a cause of concern for anyone wishing to pursue an anthroposophic spiritual science. I am referring to the fact that modern humankind is basically ignoring clearly evident powers of decline, powers that must inevitably take our present civilization to the edge of the abyss if allowed to take effect.

Surely we have to acknowledge that many things are surfacing from the subterranean depths of human nature and coming to realization in various ways; in other words there is a great deal going on at present to which many of our fellow citizens seem simply unable to pay proper attention.

It is reasonable to say that at the present time little effort is made in cultural life to take a wider point of view and pay genuine attention to the forces that shape our world. There is one school or spiritual orientation—I have characterized it a number of times over the years—that has its roots mainly amongst English-speaking peoples and is rather secretive about its work. It is, however, extraordinarily effective. A second tendency is the movement that has come together because people want to take account of the instincts of the masses, instincts that are understandable and indeed also justifiable. In its extremes this movement is represented by people who have no idea of human evolution, who know nothing of the principles that signify progress for the world. Certain conditions, however—I shall refer to these later—enable them to secure a position of authority despite their narrow-minded views and despite what is in fact a marked natural inclination for criminal activities. They are of course clever people and able to impress others, thereby making their presence strongly felt.

The third movement that has an effect in cultural life is based on particularly energetic representatives of the different confessions—confessions of all kinds—who also know very well what they want. They are at the centre of everything that usually comes under the heading of Jesuitism. Many people talk about Jesuitism and the like, but still large numbers of our fellow citizens are little inclined to pay proper attention to what is really going on here.

To get a proper idea of current events one would have to take account of a number of things. One thing to which we should especially attend, however, is connected with a fact I also mentioned in my first public lecture here.[20] It is this: people's inner state of soul today, in particular their state of mind and the structure of their thoughts, is a continuation of a very great many things that were only suitable for the medieval frame of mind. That was a great and significant way of thinking, but it is now outdated. Some people have wholeheartedly embraced medieval forms of sensibility and thought. These are the ones who hold more or less socialist views all over the globe. The ideas current among them come to expression above all in a belief in authority that is almost limitless. They kowtow before those amongst them who assume authority simply by asserting their dominance. This has made it possible for people like Lenin and Trotsky[21] to impose their tyranny on millions of people with the help of just a few thousand. That particular movement is spreading from Eastern Europe into Asia at an incredible pace. It imposes a tyranny worse than anything seen during the worst periods of Oriental tyranny.

All these things need to be considered in forming an opinion about current events. I can only give a brief outline of these things here. Basically the only opposition to these trends—and we are still thinking in terms of major forces in world history, forces shaping the world—comes from what should ideally be a truly honest, sincere and genuine spiritual-scientific movement. If we compare the interest brought to this spiritual-scientific movement with the interest those other movements have aroused within a relatively short time, and with the influence these movements have gained, we have to say that interest in this spiritual-scientific movement is as good as nil at the present time.

We do not fail to recognize of course that there are many people who affiliate themselves with this spiritual-scientific movement, or at least tell themselves that they do so. There would be an enormous difference, however, if people really took note of the intensity with which those other three movements work for the things they want to achieve, and then compared this with the intensity of interest that exists in spiritual science. People engage with the spiritual-scientific movement with only very superficial feelings: the feeling element is very weak. The other movements I have spoken of on the other hand arouse limitless intensity of feeling.

Does anyone clearly understand—and place it at the very centre of their heart and mind—that if spiritual science is to engage to any serious extent with forces that shape the world, people must first of all give due recognition and proper value to initiation knowledge, or initiation science as we call it? Initiation science today encompasses humanity's keenest interests. Many people believe they are sincerely devoted to it, yet the interest they muster is still rather superficial, and focused on all kinds of insignificant considerations.

The people I have often referred to as the real movers and shakers in the global Anglo-American movement, have initiation knowledge, though it is certainly not used for the benefit of humankind. Everything based on Jesuitism has initiation knowledge; and in its own peculiar way Leninism also possesses initiation knowledge. Leninism knows how to put things cleverly, using rational ideas produced in the head, and there is a definite reason for this. The cleverness of the human animal, the cleverness of human animal nature, comes to expression in human evolution through Leninism. Everything arising from human instincts, human selfishness, comes to interpretation in Leninism and Trotskyism in a form that on the surface seems very intelligent. The human animal seeks ascendancy, seeks to be the most intelligent of animals. As I have often stressed, it seeks to shape humanity by drawing on all the ahrimanic powers that exclude the human element, exclude all that is specifically human, and instead employ all the aptitudes that exist within the animal kingdom.

Consider—and this is something else I have often stressed—the conceit shown by humans when they invented things such as linen paper, paper made from wood or the like; in short, paper of any kind. Well, wasps and similar creatures invented this a great deal earlier, building their nests from the same materials as those from which we make paper. There you have human cleverness inhabiting animal nature. If you now take all the cleverness of this kind that exists within the whole animal kingdom, and imagine ahrimanic powers drawing on and bringing it to life in human heads, in the heads of people who follow only their egotistical instincts, you can see that it may be true to say that Lenin, Trotsky and others are the tools of those ahrimanic powers. That is an ahrimanic initiation. It belongs to a sphere of existence different from our own. It is, however, an initiation that also holds the potential for eradicating human civilization on earth, getting rid of all that has evolved as human civilization.

We are therefore dealing with three schools of initiation. Two lie within the plane of human evolution and one descends below that plane, though it is an initiation of tremendous willpower, almost unlimited willpower. The only thing that can bring order into all these developments, setting a goal that is worthy to be called human, is contained within genuine spiritual knowledge. A true goal and genuine sincerity will, however, only come from this spiritual science if we make it into something that fully involves us, at the same time observing how much empty chatter, how much conceit and inner egotism comes to expression in so much of what is usually said of it, despite being people's honest opinions. These things cannot be left unsaid. On the contrary, we need to discuss them repeatedly. How else can we hope to give souls the power today that is needed to prevent civilization's utter decline?

Let me take a few minutes to give you a very tangible picture. Just a short time ago I read the following in a newspaper:

> Religion is a fantasy that arises in human heads as a reflexive response to the way they relate to each other and to nature. It is doomed to die a natural death in the triumphant progress of a clear, scientific and naturalistic interpretation of the truth that will develop hand-in-hand with the planned rebuilding of society.[22]

Considering how deeply asleep souls are in the present age, we may ask ourselves how many people reading this kind of thing in a newspaper article pull up short, as though stung by a viper, at the most terrible symptom expressed in these lines. People do not reflect on what would happen on this earth if these words were ever realized:

> Religion is a fantasy that arises in human heads as a reflexive response to the way they relate to each other and to nature. It is doomed to die a natural death in the triumphant progress of a clear, scientific and naturalistic interpretation of the truth that will develop hand-in-hand with the planned rebuilding of society.

'Religion' does not refer here to some confession or another, nor to some religious movement that one might quite rightly consider to be wrong, nor merely to religion in the narrower sense, but to all ethics and morality. If the thoughts expressed in these lines were to come true, human society across the globe would very rapidly degenerate into herds of animals, albeit animals capable of very sophisticated thought. If the trend that is growing in Eastern Europe and spreading into Asia at an incredible pace does not meet with opposition, civilization will be doomed. The ideals expressed in those lines would then become reality.

In the light of such impulses in world history I do not think it is justifiable for people in certain places to pursue mystical small-talk within closed circles, small-talk that, counter to my intentions, has eventually also arisen in anthroposophically oriented spiritual science. Some people even consider this ideal! I do not think it is right to continue like this, and to disregard what the wider interests of humanity on earth require of us. We must have the will to consider humanity's greater interests without bias. We must give due and serious attention to certain basic principles—not merely in theory, using our intellect, but instinctively. Those principles have been obscured by all the confessions in Europe and America and the intention is to obscure them still further.

We know about the virulent propaganda campaign being launched against anthroposophically oriented spiritual science; you can hear the bullets whistling all around. Wherever opposition arises, it would be a shame to indulge in the harmful illusion that we may ever hope to achieve anything by converting people who, after all, are the very agents of something that belongs to the past. We cannot and must not be opportunists or seek compromise. This should, as it were, be our special meditation every morning. Well-meaning people have said we should simply try to explain to people somehow or other how we are endeavouring to bring the Christ Mystery to the world. But the more we do this, the more bullets whistle around our ears from certain quarters. The idea that humanity should today gain true understanding of the Christ Mystery is completely at odds with some Catholic or Protestant groups, for instance. It is not in their interest that the true Mystery of Christ comes to be known; all they want is to hold on to the old ideas. If we had some kind of strange and peculiar creed concerning Christ they would treat us as a harmless sect, as odd characters, and not fight us with the intensity we have come to experience. Within the two cultural movements, quite apart from the third, there are, however, quite a number of people who know that our aim is to speak of the Christ Mystery in truth, and to address issues relating to the social order based on the threefold principle. This makes them sit up and listen; it makes them say: 'It would take the ground away from under our feet if we were to embrace the truth; let us therefore vow to destroy it.' People do not fight us because we are in error, they fight us because in certain quarters they realize that we seek the truth. There is no point in speaking in any other way about some of the things that happen today. The spiritual movement I am speaking for here has a profound interest in absolute clarity, particularly also clarity of thought.

Try to recall some of the things I have discussed before. What, after all, do we need to recognize in relation to the primary needs of humanity today? The important thing is that our powers of thought—everything we have in the way of our ability to form ideas, apart from sense faculties—are a legacy from our life before birth or life before conception. Our whole capacity to think is one we brought into the physical world when we were born; we brought it with us from the life we had before we were born. All the thoughts we elaborate whilst we are in our physical bodies are powers that govern our whole being between our last death and the birth process that brought us into our present life on earth. When we are thinking here and now, the powers of thought we use, not the thoughts themselves, are a shadow image of something that was at work before we were born or conceived.

Try to think of what we nowadays call natural forces, of what acts in lightning and thunder, in the movement of waves, in the way clouds are formed, in the rising and setting of the sun, in wind and rain, in the way plants grow from the ground, in the way animals are conceived and born and grow. Think of all the natural processes you see all around you; then think of them merely as a picture, not the reality. So, please, imagine these natural forces casting their shadow somewhere, and these shadows being contained in a vessel and presenting themselves to us as pictures. The relationship that exists between present natural phenomena and the reality that underlies them is similar to the relationship between life before birth and our faculties of thought in our present life on earth. Just imagine that you have here everything that happens to your soul between death and rebirth—I am describing it schematically—and then it casts a shadow; a shadow arises of everything you have there and this shadow becomes the content of your head, the content of your thoughts; it is your faculty of thought. What you are thinking now, those are the forces active before you were born. That is 'nature' in the spiritual world, if I may put it in such a paradoxical way. The evolution of humankind cannot progress unless we become aware that when we are thinking, the existence we had before birth influences our faculties of thought. Having entered into my present earth life, I am continuing the life I had before birth when I think.

Who most strongly opposes this idea? The greatest opposition comes from religious confessions which maintain more or less the following: 'A human child is born. It pleases two people, a male and a female individual on this earth, to come together, and God creates a soul in the spiritual world, a soul which then connects with what is created between two people in the act of begetting. That is how the human individual comes into being.' This is, of course, very different from what I have just been saying. And it is the lifeblood of religious denominations in our modern civilization. They all teach that when two people conceive a child, the spirit very kindly creates a soul up above, a fresh new soul; it is then sent down to unite with the physical body which has been created, and something new has come into existence. To whom do these beliefs address themselves? They address themselves to terribly egotistical individuals who simply cannot bear the thought of being extinguished when they die. Yet they are able to bear the thought—for they have got used to it over the centuries, indeed soon it will be millennia—that it pleases God to create fresh souls for human beings procreated here on earth. What their egotism does not allow them to accept is the thought that death puts an end to it all.

Of course you all know the nature of life after death. I do not need to go into it here. But let us turn our attention to something quite different. Preachers in their pulpits always need to assume that they are speaking to people who cannot bear the thought of death being the ultimate end. The wisdom they must pour down from their pulpits—irrespective of the particular creed followed by the people who sit there below them—must make it clear to them—I mean unclear, of course—what happens after death. They have to choose words most liable to excite the egotism of people; they have to utter phrases that are fully in accord with the egotism in people's souls.

Let us think what would happen for instance—to give a particular example—if someone were in all seriousness to question certain Roman Catholic tenets, say the dogma that when two people conceive a child it must please God to send a freshly made soul down to them. What would happen if this were to be questioned? Someone exploring the whole issue without prejudice would find that it has nothing whatsoever to do with anything to be found in the true Christian faith. They would find that during the Middle Ages the teachings of Aristotle infiltrated theology and that Aristotle represented these ideas, based on misunderstood Platonic ideas, which state that a fresh soul is created for every newly generated human body and unites with it. Something taken for granted as a fundamental tenet in Christian beliefs in fact has nothing to do with Christianity but is an Aristotelian principle.[23]

Here's another example. One element in religious beliefs is the dogma of eternal punishment in hell. Again, entirely an Aristotelian thought. Aristotle assumed that once a soul had been created, lived on earth and then entered the spiritual world, there was nothing it could do in the spiritual world, as he saw it, but look back for all eternity on what it had done during its one and only life on earth. Aristotle imagined that a fresh soul was created for every child, that this soul lived on earth until the individual died and then for all eternity occupied itself with the contemplation of what had happened during one life on earth. If someone had committed murder, they would have to look back on this for ever. That is where the dogma of eternal punishment in hell originated. It is a purely Aristotelian concept.

Just think if the truth were to become known instead of Aristotelian thoughts presented as Christian dogma. The people wishing to represent such Aristotelian ideas masquerading as Christian dogma would be scared out of their wits that people might find out about this, that they might discover that their priests were not teaching Christian ideas from their pulpits, but Aristotelian ideas which had infiltrated Christian teachings.

Christian beliefs also contain an infinite number of ideas derived from gnostic teachings. The Roman Catholic celebration of the Mass contains an enormous amount that originated from the Egyptian mysteries. Many of the rites of the Catholic Church—and the Protestant, too, in many respects—contain things whose origin must be sought in all kinds of Oriental religions. Their paramount concern is that no one should find out where these things come from. What do they feel compelled to do? They have to resort to slander! They have to say that the people who present the truth today are plagiarists borrowing from Oriental and gnostic teachings and so on. 'Traubism' is the order of the day. They come up with learned calumnies like those presented by the clergyman Professor Traub[24] and all the people who parrot him. Why do people do such things? Because the truth is emerging and they all have an interest in not letting it emerge. People will go on saying that what we are doing is borrowed from some other source. They will provoke in their congregations an opposition to gnosis and things that are part of the very fibre of their own souls because they do not want it to come to light in its true form. Gnosis—one is supposed to say—is something terrible, something dreadful. Then people will ignore it, being afraid of it, and the preachers can talk about things that in fact have their origin in gnosis. It is these very preachers in fact, rather than those who speak of anthroposophically oriented spiritual science, who derive what they say from gnostic texts. Such preachers are most afraid of the reality of the soul's pre-existence, a life of the soul before birth and conception, the soul's roots in the spiritual world through all the ages as spiritual science alone can disclose to humanity. For if the truth were to become known, there would be no room any more for the blasphemy that the gods are obliged to send a freshly created soul from the spiritual world to unite with every single human body. All these things have their origin of course in an ever-growing desire for power. Behind it all are thoughts of power. It is possible to put tremendous energies into such desire for power simply by following certain precepts.

What is going on in Dornach at the moment, for instance? Almost everywhere in Switzerland articles on anthroposophy are being published, not one sentence of which is true.[25] The whole campaign started when an article appeared that contained twenty-three lies. For weeks now, article on article has picked up on those twenty-three lies; they have appeared almost everywhere in the Catholic Press in Switzerland and not a single sentence is true. Why is this happening? It happens because a certain state of mind is invoked in the many followers of these people, one where it is no longer possible to tell the difference between truth and falsehood.

Think of all the efforts we employ in anthroposophically oriented spiritual science to form sufficiently clear ideas; for instance, about the degree to which dreams may or may not reflect the truth. As human beings we cannot immediately distinguish truth from falsehood when something appears in the course of a dream. The same state of mind arises for a congregation when they are told lies by people who know that those lies will be believed. The soul is brought into a particular state, a mood, by those lies and becomes the willing tool of those desiring power. It is easiest to get people into your power by planting illusions in their unsuspecting minds. Articles full of lies are systematically circulated with the intention of creating the kind of mood that can be engendered with lies. That will be the inevitable consequence of the probabilism which the Jesuits have been teaching for a long time. It is merely a final consequence.

It is of course difficult to rouse modern souls from their general torpor to stand up against such people. The day before I left we were forced to arrange for a lecture—for we must fight, of course, even if we do not want to, against the lies that reach Dornach. Dr Boos, one of the most courageous of our young protagonists, called on everyone who had anything to say on the subject of the lecture to join in the discussion—it was a public lecture, of course. When no one came forward he openly and publicly declared the cleric who had first written those twenty-three scurrilous lies, a priest called Arnet in Reinach, to be unworthy of his priestly calling.

There is no other way forward. And even when this had been said, only one individual stood up among those present, a teacher, shaking in his boots if I may put it like that, and said: 'Just wait. There are more articles to come, and in the end you will see!' Well, all I could say was that twenty-three lies began all this, and the end of it all, when the truth about those twenty-three lies emerges, will doubtless never arrive, not if we had to wait till the end of the world. Not the least attempt has been made in everything published so far—and a respectable number of articles have already appeared—to investigate those twenty-three lies.

Other things have been tried, using a strange logic. The pamphlet by the Tübingen speaker played a big role—but the people who bring professor Traub's pamphlet into play in their articles have not properly understood what he said. They will write that this man Steiner is borrowing from all kinds of ancient writings, from the Upanishads, the Egyptian Isis Mysteries and the 'Akashic Records'—well, I suppose the typesetter may have put that in, but on the other hand the clerical gentleman may have done so. I therefore said that it was not really my concern to correct printers' errors, but that it surely is a strange way to interpret Traub's pamphlet if not even Traub says anything as stupid as that the Akashic Records are volumes found on library shelves, from which someone has plagiarized to create anthroposophically oriented spiritual science.

Our attackers have also gained support among liberal thinkers. Dr Boos wrote a powerful piece in a liberal paper to highlight the deliberate untruth of this, since the writer must have known there were no Akashic Records in his library. He could not possibly have them in his library and so he ought to have known; he must have written a deliberate untruth. What did the person concerned do? He replied that Dr Boos was evading the issue, as it was self-evident that the typesetter must have been responsible for the 'Akashic Records' error and not he himself. In his view the kind of sophistry that made authors themselves responsible for that kind of printing error merely revealed the ill-will of such people.

Well, you see the kind of mentality one is dealing with. But do not underestimate it! You have to realize that it is going to be a hard fight, one that increasingly resorts to such practices. The aim is to prevent people from finding out about things I have been saying, such as the following. In the medical course I first said that in making serious efforts to draw the spiritual laws of this world from modern life, trying to fathom the deeper secrets of human nature through observation of contemporary life, and then also finding them written in ancient works—albeit arising from a culture that was more instinctive and atavistic—that one feels very humble to discern the greatness of the instinctive, atavistic intellect that human beings once possessed. This has been lost and must now be rediscovered. I spoke these words in awareness that knowledge which today has to be sought within life was once instinctive wisdom given to humankind. Much of that ancient wisdom has of course survived in the religious beliefs, though it has been corrupted. Yet the people professing those beliefs want to make humankind fear that ancient wisdom, and when they talk about it say more or less the following: 'Those dreadful people who pursue anthroposophy today are borrowing everything from ancient teachings.' If they investigated the matter they would find that the spiritual science offered to humankind in anthroposophy is very different from anything ever borrowed from anywhere, from the Upanishads or elsewhere. So we had to borrow indeed from that ancient tome called the Akashic Records! Thus to prevent people getting wind of something that belongs to the present age, our enemies fire off their bullets from all directions.

Let us be clear about one thing. You may feel tempted now and then to stress the good points of one thing or another. The ever closer alliance between Jesuitism and the Social Democrats is something entirely natural. There is nothing unnatural about it. The Social Democrats are equipped with the same kind of ideas as the Jesuits, only they reverse them, see them from the reverse aspect. The sense of the human being's 'eternal nature', however, is entirely different in so far as it has become a tenet of egotism. It is restored to its true form when the pre-existence concept—that of a human soul having a life before birth, or before conception—once again becomes the active moral principle. The knives will come out to fight this idea. We shall only be able to progress in the world if in the first place truth has inner power. This inner power can only be effective, however, if in the second place people have the courage, however few they may be in number, to carry this truth in their souls, carry it in their souls in all seriousness, uprightness and honesty and without compromise. It is useless for us to overlook the tremendous difference which exists between true Christianity and Catholic and Protestant Aristotelianism, which holds that souls are created for bodies as and when they arise through procreation. We must not play down this difference. If we do play it down we will not even notice where the idea of power, the desire for power, truly originates.

I find myself referring again and again to the pastoral letter issued by a Roman Catholic bishop.[13] This document states that the faithful must regard their priest as ranking higher than God and Christ, for each time the priest performs the consecration at the altar, Christ is forced to be present at that altar, to be present in the bread and the wine which is His body and His blood. The priest therefore has greater power in the universe than a god. That is what it says in an actual pastoral letter, which has also been requoted in many other pastoral letters. Now you may ask if that is consistent with the abolition of the spirit by the Council of Constantinople[26] in 869 AD. The answer is yes. A Roman Catholic saying that God is more powerful than a priest only says so because people will not, after all, accept the opposite view nowadays. People are so asleep in their souls that they never even wonder what the person[27] writing to Moleschott really meant when he had the courage to say that a criminal, a liar, a murderer is a moral person in so far as he gives expression to the entirety of his predisposition; and is an immoral person if he does not bring to expression what he has in him, for this would impose restraints on his individuality; accordingly, an inclination to murder is just as valid as other inclinations are. Modern souls do not have the courage to recognize that if scientists continue to teach the kind of underlying outlook that they have been teaching, the inevitable conclusion simply has to be that criminals, murderers, are just as good as someone who seeks to act ethically. People merely lack the courage to admit this. When materialism had its flowering, at the time when people like Vogt, Moleschott and Buechner,[28] all of them courageous men, were publishing their writings, such things were acknowledged. The present age is too cowardly, however, to admit such a thing. Nor is there sufficient courage in the sleeping souls of today to admit that the intrinsic spirit of creeds and religious confessions is that a priest is indeed more powerful than a god.

The school of thought represented by anthroposophic spiritual science must above all work towards clear thinking in every respect. Its message cannot be grasped if thoughts are unclear, it cannot be grasped in a vague and vaporous mysticism but only with crystal-clear thoughts, thoughts which, in my *Philosophy of Freedom*,[29] I have tried to show are the starting point for genuine human freedom.

We can continue our discussion of the subject when I am able to speak to you again. I hope this will be soon.

Lecture 5

STUTTGART, 24 JUNE 1920

TODAY'S meeting provides a further opportunity for me to speak to you as friends of the anthroposophic movement, before I leave. It is a heartfelt concern of mine to speak of things that really need to be discussed. It is possible that most of what I have to say today is a repetition of things that have been discussed on a number of occasions from all kinds of different angles, things now also referred to in public lectures. There are reasons, however, why it is necessary for us to consider some of them once again today.

I have often stressed that a sufficient number of people need to fully understand the following. To prevent the galloping decline into which we have fallen in the civilized world from leading to utter ruin, certain impulses must be introduced into modern civilization that can only arise from the fullest scope of spiritual-scientific insights.

Materialism has come to Europe over the last three or four centuries, reaching its peak in the nineteenth century and then surging on into the twentieth. It has a peculiarity that seems paradoxical, particularly if one fails to realize the true causes. The peculiar thing about materialism is that it is incapable of recognizing the material world as it really is. I believe I have already given you an instance of this. The materialistic way of thinking has in more recent times given rise to an idea that a great many people believe, namely that the heart is a kind of pump that pumps blood through the organism. All sorts of versions of this idea of the human heart being a pump surface today. The facts are rather different, however, and should be seen like this: the whole of our rhythmical circulatory system is something alive.

It cannot be compared with a system like channels with water flowing through them, water kept circulating with the aid of a pump. Our rhythmic circulatory system, our blood system, is something alive. It is kept alive by a number of factors, the major factors being breathing, hunger, thirst and so on. These clearly function at the level of soul and spirit. Our blood system is set in motion by entirely primary causes, and the movement of the heart arises when this spiritual principle enters the rhythm of the blood. The rhythm of the blood is the primary, living principle, and the heart takes up this rhythm, is taken up by it. The facts are therefore the very opposite of what every professor of physiology teaches today, which in consequence is dinned into people's heads at school, impressed upon them indeed from their earliest childhood.

It therefore has to be said that materialism has not succeeded in gaining real understanding of the physical processes in the human organism at work in the heart. It completely fails to understand the material aspect in particular. This is just one of many examples. Material things in particular remain completely unexplained by materialism. The heart is not a pump. It is something we might regard more as a sense organ incorporated within the human organism to give human individuals a kind of subconscious perception of their circulation, just as the eye perceives colour in the world outside. Basically the heart is a sense organ within the circulatory system, yet exactly the opposite is taught nowadays.

This would appear to be an example of limited relevance. I can imagine some philistine thinking there is little harm done if people have entirely the wrong idea about the nature of the human heart, though such a person will no doubt accept that it would be worrying if physicians harbour misconceptions about the heart. But such misconceptions also extend their influence to other things. Everything is connected with everything else in life, and because of this humankind is absolutely full of wrong ideas, completely upside-down ideas. Attachment to misconceptions can be seen, if one takes it seriously, to cause real havoc in our thinking processes. It certainly does so. Thinking is ruined by having such things dinned into us: we have become used to thinking that things are the opposite of what they really are. That is why we never acquire the habit of steady, purposeful thinking. How can our thinking grow purposeful in social life, for example, if in areas where truth should be sought above all else we are in fact departing from it?

You see, some things that are important to know are a closed book for people today. When the human organism is studied in mainstream institutes nowadays, in physiological and biological laboratories, in hospitals and similar institutions, the brain for instance is examined by analysing it bit by bit as it presents itself to the eye. The liver is examined by the same kind of analysis. In doing so, people never consider one thing that is absolutely essential if one wishes to understand the human being: the whole head organization as it exists today, and everything it governs, is entirely different from the rest of the human organism.

Let me show you what underlies this. You can draw it like this. I intend to lead up gradually to what I really want to say. You can say that the human being has two organs of sense perception, and the direction in which they perceive is approximately like this [see (a) in the diagram]. Two other directions in which we perceive show a certain relationship to these. In diagrammatic form I would draw them like this (b):

The human being thus perceives in four directions, as shown in the diagram.

I deliberately did not tell you where these organs are to be found in the human organism. If I draw nothing but two arrows to indicate direction (a) here, which we extend, as it were, to perceive, and two others here, (b), through which we perceive side-on, it makes no difference at all if these are the directions in which feeling and

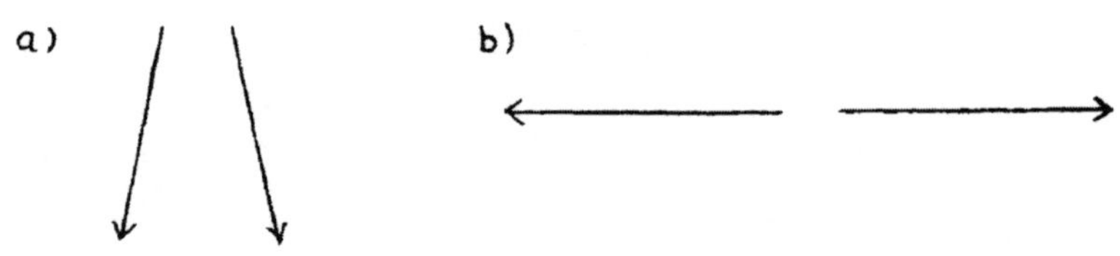

sensation pass through my legs and these where they pass through my arms. Here we have something that is in accord. I perceive my own gravity, as it were, I stand with my two feet on the ground. I really perceive something. And I also perceive something when I stretch out my hand, stretch out my arm, even if I do not actually touch anything. I can draw it like this (a). The same drawing can also stand for something different. Imagine this is the horizontal plane. The two arrows could represent the two visual axes; I could draw the two visual axes like this. And these arrows (b) could indicate the directions of my ears, my auditory perception. The same diagram would serve to indicate perception by the eyes and ears. In the one case I have the whole organism within the head, though the plane has turned through a 90° angle, in the other case I have it within the rest of the organism. There is a higher point of view where both are the same. Our two legs are merely directions in which we perceive, that have become flesh. The same directions exist in a less physical form where they extend from the brain through the eyes to perceive colour. Elsewhere we perceive gravity and everything connected with it. We see our weight and we step upon colour, we could say, if we were to reverse the two things, but do this in an entirely organic fashion. I hear the blackboard chalk, I touch a C or C sharp that resounds. The difference is merely one of degree. In the head everything has gone through a 90° angle and is less physical; the other is in the vertical plane, and is physical. In the final instance both are the same. It is only that I am aware of what my eyes step on as colours, my ears touch as sounds; I know about it, it is part of my ordinary conscious life. Everything my legs see of the conditions of gravity, and all the other kinds of things that my arms hear—all these reside in the subconscious sphere. Conditions belonging to the cosmic sphere are present in the subconscious. With the whole of my subconscious I have knowledge of the cosmic sphere, knowledge of the way the earth relates to other bodies in the universe, knowledge of the universal contexts of gravity. I hear the music of the spheres with my arms and not with my ears.

Thus we may say that we have what can be called a lower organism, with subconscious cosmic awareness, and we have a head with earthly awareness; this however is a 'conscious' awareness. The organization of the whole human being is based on these differences. Our outer form and configuration depends entirely on them. You know that the head we carry today is the transformed body of our previous incarnation, our previous earth life, and that the rest of our present organism will be the head in our next life. The head, then, is the rest of the organism which has undergone a transformation. It is more perfect, more finished in a way. As a result, legs have become fine visual threads extending from the eye and stepping upon the colours in a very lively way. The arms of our former life have become so ethereal that they now extend from our ears and touch the sounds we hear.

These are tangible facts about the human being. It does not get people anywhere to know about repeated earth lives and so on. Those after all are dogmas and it makes no difference if you have the dogmas of the Catholic or Protestant Church or the dogma of repeated earth lives. Real thinking only starts when you enter into tangible realities, when you come to realize that looking at the human head you are looking at the transformed body of your previous earth life, and that the head you had then was the transformed body of the preceding life—you must imagine it without the head, of course. The head you see now is the transformed organism of the last life lived on earth. The rest of the organism as you see it now will be the head in the next life. Then the arms will have metamorphosed and become ears, and the legs will have become eyes. We must look at the physical world and understand it in its transformed non-physical form, our intellect must illumine the material world in this way. Then at last we shall have what humankind is so much in need of today. Once the human mind has been organized so that it no longer produces the kind of folly that has been proposed as social theory, particularly in the second half of the nineteenth century, human beings will indeed be ready to develop social ideas that can be put into effect in this world. It is necessary to gain a thorough understanding of this today. People today should recognize, as a matter of serious concern, that the science which is so highly respected, and so widely disseminated, must be replaced by something else. This is indispensable.

It is nonsense, as I also said recently in a public lecture,[30] to talk about establishing general adult education classes where things are studied in the same way as at ordinary universities. It is the work done at the universities that has brought us to these disastrous situations, founded on the materialistic view of a few leading personalities. This outlook is now to be presented to the masses; that is, millions are now to be dragged into the disasters created by the false leads dictated by a few. Something that proved useless for a few is now to be spread among many. It is not as easy as that, however. Popular education cannot be introduced simply by teaching outside the universities what until now has been cultivated within them. This would mean teaching something altogether unsuitable for human beings. This may sound radical, but it is absolutely essential that it is fully understood if there is to be even the least hope of the decline being halted and something new and positive developing.

These are things one wishes one could speak of in words that truly reach people's hearts. These tangible truths must reach as many hearts as possible. It was therefore important to me to point out in my public lectures that something has been achieved in the Waldorf School, that anthroposophy has positively influenced some aspects of history lessons. I was also able to refer to the teaching of anthropology in Class 5. There, too, anthroposophy has proven fruitful. Not that one would teach anthroposophy to the children—we would never think of doing such a thing—but lessons come to life if anthroposophy is the foundation, if the inspiration of anthroposophy is present in what we teach. This brings children's souls to life; they are quite different when this influence is there. It would be taking the easy way simply to teach anthroposophy in our schools. No, that is not what we are about, but rather we use anthroposophy to enliven the subject matter. It will of course be necessary for anthroposophy to come alive in ourselves first of all, and that is something that really proves infinitely difficult, to bring anthroposophy to life in human beings. But the potential is there today for all kinds of disciplines, not only science but all kinds of areas of life, to benefit fully from the living quality which anthroposophy can engender.

That is a general way of looking at it. Let me go on to something specific, so that you can see the things we are considering in their proper context.

Marxist philosophy, Marxist views are widespread today. They have their most radical expression in Leninism and Trotskyism, which are destroying the world. A view of history known as 'historical materialism' plays a great role in Marxist philosophy, particularly the dogma of the fundamental importance of the modes and circumstances of production. Millions of proletarians have accepted this dogma according to which tradition, law, science, religion and so on are like smoke, like an ideology rising from the modes and circumstances of production—you will find further details in my book *Towards Social Renewal*;[31] the modes and circumstances of production are seen as the only reality on which to base one's view of history.

In the past it was very important to me—connected with the feeling I have that I was really able to achieve something at the Worker's Education Institute in Berlin[32] and create a potential foundation there—to speak to proletarian circles about this idea of the modes and circumstances of production as the only determining factor in human evolution. My aim therefore was not to teach historical materialism but to disseminate the truth. That was of course also why I was thrown out, for it offended those in charge at the time, just as much as the idea of a threefold social order offends people today. Authoritarian thinking and belief in authority were and still are as great in the socialist movement as in the Catholic Church.

What really matters is to gain a clear understanding of social relations in the world. Real understanding of the natural threefold order of the human organism, of the way the human organism is an organism of nerves and senses, a rhythmical organism and a metabolic organism, as shown in my book *Riddles of the Soul*,[33] leads to a way of thinking that can also apply to social life. Those with little understanding of such a thing will think I am using an *analogy*, applying the threefold form of human body to the social organism. This is nonsense of course. Analogy is not the method used in *Towards Social Renewal*. All I said was that if people succeeded in freeing their thinking from the straitjacket imposed on it by modern scholarship and, particularly, public opinion, they would liberate their thinking and enable it to form sensible ideas about social issues. The kind of thinking that regards the human brain in the same terms as the liver, examining everything as though it were of the same substance, will never come to sensible conclusions.

If we were merely to create outward analogies we might say: The social organism is threefold by nature and so is the human organism. The head is the organ of mind and intellect; it should therefore be compared with the cultural and intellectual life in the threefold organism. The rhythmic system establishes harmony between different functions in the action of the heart, in respiration—that would be the rights sphere in the social organism. Metabolism, the most physical, material aspect—something mystics tend to look down on to some extent, though apparently they also have to eat and drink—would be compared to the sphere of economics.

But this is not so! I have repeatedly pointed out on other occasions that in reality things are very different from how mere analogies would seem to make them. It cannot be said, for instance, that summer is comparable to the waking state for the earth and winter to a state of sleep. The reality is different. In summer the earth is asleep, in winter it is awake. I have gone into this in detail.

The same applies if we compare the *realities* of the social and the human organism. The economic sphere of the social organism must actually be compared to the activities of the human head. As to the sphere of rights, since this is the mediating and middle element, it is indeed true to compare it with rhythmic activities in the human organism. The life of mind and intellect, cultural life, however, has to be compared with the metabolism. This means that economic life has to be compared with the organs of mind and spirit; and the cultural and intellectual sphere of the social organism with the metabolic organs. There is no way round this. Economic life is the head of the social organism; cultural life is the stomach, liver and spleen of the social organism, not of the individual human being. It is of course too much of an effort for anyone whose thinking is straitjacketed to distinguish between the life of society and the life of an individual person.

Again the essential point is that spiritual science prepares us to see things as they really are and not to produce neat analogies and elaborate symbolism. We will then arrive at important conclusions. We will find, for example, that, in this case, if economic life really is the head of the social organism, it will have to depend on, live from, the rest of the organism, just as the head does in the human organism. In that case we cannot say morality, religious life and the search for knowledge are ideological elements *arising from* economic life. On the contrary, economic life depends upon cultural life, on the metabolism of the social organism, just as the human head depends on respiration, on stomach, liver and spleen. We then come to see that economic life arises out of cultural and religious life. If we did not have a stomach we could not have a head. Of course we also could not have a stomach if we did not have a head, but it is the head after all that is fed by the stomach, and in the same way economic life is fed by cultural life and not the other way round. The socialist theories that now threaten to spread throughout the civilized world are therefore quite erroneous, a dreadful superstition. No one has thought to look for the truth in recent centuries; based only on emotions, everyone has been promulgating the kind of truth their social class and point of view suggested to them. Only now are people realizing that it is complete illusion to see historical evolution as the outcome of the modes and circumstances of production, and instead are beginning to compare actual realities rather than speaking in analogies. Realistically speaking, in the same way as the head will suffer if the stomach is ruined, there can be no sound metabolism in the social organism, and economic life must fall into decline, if morality, religious life and intelligent thought are undermined. Nothing at all depends on economic life; everything depends first and foremost on outlooks, ideas, the cultural life of humankind.

The head is always dying—I have spoken of this in other lectures—and we only sustain the head organism by virtue of its constant dying, a death against which the rest of the organism continually rebels. The same applies in the sphere of economics. Economic life is constantly bringing death and decay into the progress of history; rather than generating everything else it brings about the death of everything. This element of death constantly has to be counterbalanced by what the cultural organism is able to produce. The truth is therefore exactly the reverse of what is assumed. Anyone speaking in materialistic terms and saying economic life is the basis for progress is not speaking the truth. The truth is that economic life is the basis of something that is always gradually dying, and the mind and spirit have to redress this dying process. To proceed the way people are now proceeding in Russia is to help the world to its death. The only possible outcome of proceeding in this way is to help the world die, for the simple reason that the laws of death are inherent in the things that are being done there.

You can see the eminent social importance of these things. We have now been working in anthroposophy for twenty years, and all the time I have tried to make it utterly clear and apparent in all kinds of lectures that what matters to us is not the cultivation of a philosophy full of inner self-gratification, a kind of spiritual snobbery, but developing an impulse that is essential, and so urgently needed today.

I wanted to present this to you again today in a slightly different form in connection with a number of things that can help us understand the essential nature of the human being. It is important that those who call themselves friends of the anthroposophic movement clearly perceive the connection between this movement and other events occurring in our midst.

The ideas put forward by myself and other friends are often seriously distorted. It is therefore difficult to speak freely to a large audience, even if it be an anthroposophic gathering. Since there is no immediate opportunity, however, to discuss these vital things at a more intimate level, let me draw your attention to a few matters. We must be aware, particularly here in Stuttgart, that the anthroposophic movement that has existed now for twenty years has in fact reached a new stage. If we are serious about the movement, this means we accept the duty to allow and adapt to this change. You must properly understand that because our friends Molt, Kühn, Unger, Leinhas[34] and others have attempted to take the anthroposophic approach to its practical conclusion, something has happened that concerns us all. It concerns us all and we must take account of it in everything we say and do. The fact is—and let us be very clear about this—that up to that point the anthroposophic movement was a spiritual current in the world. Such things pursue their paths. However objectionable this is, cliques and closed groups may form in the pursuit of personal and all sorts of other interests; a spiritual movement may even proceed by the mal-'agency' of privy councillors like Max Seiling.[35] One does of course have to handle such things properly, but as long as they are a purely spiritual or cultural movement they can be ignored. Now, however, three things have grown out of this spiritual movement of ours.

The first followed the appeal I made last year.[36] It now forms part of the struggling threefold movement, the Association for a Threefold Social Organism. This has not yet been able to get anywhere near the real objectives. The appeal has in a sense met with rejection, and it would be a good thing to be fully aware of this rejection, recognize that only a very little of what was intended has come to fruition.

This does of course mean that many requests come toward me. The idea has come up in Dornach, for example, of issuing a further appeal that would make it known internationally what Dornach signifies for the world. I had to explain to our friends that in the ordinary external life that is now heading for collapse, appeal usually follows appeal, programme on programme. We cannot do this if we work out of anthroposophy. It is important to know that it is not actually at all healthy to undertake something that is not realized. It is important to make a careful assessment of the chances of success, and not just do what comes to mind but only the things that have a chance of success. This is why I then said—it is important and I must ask you to consider it carefully—that I would not dream of making a similar appeal again, for what has happened to the first appeal should not happen a second time. I allowed the appeal for a Cultural Council[37] to be issued, for that was not my own undertaking; but we must be very clear that things are getting a great deal more serious than people are inclined to think if something like the anthroposophic movement stands behind them.

Three things have now in a sense evolved out of the anthroposophic movement, each of them quite distinct:

A threefold social order as expressed in that appeal—we will have to work at it, for it partly meets with rejection; secondly the Waldorf School;[38] thirdly the financial, commercial and industrial enterprise called '*Der Kommende Tag*' (The Coming Day).[39]

Coming to Stuttgart in the past, when we only had the anthroposophic movement—I am referring only to Stuttgart—I would spend three or four days here, and you know how many personal interviews I managed. These things had some effect, as is now becoming apparent. It was not without significance that whatever had happened in the meantime—people will understand what I mean if they want to—could be put to rights again in those personal interviews. Events could then proceed until the next time. Now following those outer developments, the position has changed so that I have to attend meetings from morning till night, and indeed well into the night, and there is no question of continuing as we had become accustomed to when we were only an anthroposophic movement. Many people feel irritated that things are no longer as they were. But we have to look at all the changes and really say to ourselves: Things have changed since spring of last year and we will have to take account of this.

The situation cannot remain as it is, and a united effort is needed to ensure that it does not. It cannot remain as it is because everything that is done—be it for the Waldorf School or the *Kommende Tag*—has its basis in spiritual work. Without the spiritual work that has been done and must continue to be done there is no point to it all. The spiritual work must give form, vigour and content to the whole. To continue the way we are going would mean that the institutions which have now been established would swallow up the original spiritual movement. We would deprive it of its original foundations. Nothing growing out of the anthroposophic movement should be allowed to swallow up the movement as such.

You see, these are serious matters we have to discuss today, and I think at least some of you will understand what I mean. Things will not change, however, unless we affirm the reality of the anthroposophic work that has been done for many years, for decades. This work must be seen as something real.

I would ask you also to consider the following. There is much conflict in the world, but where is most of this conflict to be found? It takes a certain form and people fail to notice, but most of it takes place in the sphere of spiritual life. There is no end to the conflict within what we call the anthroposophic movement, for example. When our movement evolved out of older practices—as you know it was necessary to start from these—that is, when many people familiar with the old theosophical practices joined our movement, I had the feeling that a gentleman, who at the time was particularly vehement in his defence of our movement, would very soon be in conflict with a great many other people. Conflict is likely to be particularly bad in this sphere. In fact I always made it quite clear that the gentleman in question, a theosophist of the purest water, would not only conflict with others, but that his right side and his left would wrestle desperately with each other. People will find that the left side of this individual engages in the most dreadful dispute with his right side.

It will of course be necessary to develop the other, balancing pole, where conflicts that constantly arise are overcome. Such conflicts are due to the very nature of spiritual movements, because they all aim to develop the human individuality. The other pole, that of human understanding, must be there as well; at this pole of human understanding it is possible to engage with and understand a human individual, to enter deeply into another person's living impulses, and so on. It must be possible for the *Kommende Tag* and the Waldorf School we are now running to receive a sound moral basis from the anthroposophic movement here in Stuttgart, the moral basis that is the work of decades, or at least should have been. That has to be the foundation, for it is the only way in which we can progress and restore the balance between a life consisting of meetings and the necessary spiritual work, which after all should be our foundation. We cannot achieve this, of course, if, as one is continually told, dreadful things have been happening again, that someone is continually causing trouble, someone is upsetting everyone else. Well, this may be so. To date—and on this visit such things have come up again countless times—I have not been able to pursue them to the point where the second person involved, when approached, told the same story as the first. When it came to the fifth or sixth person, I would hear the absolute opposite of what the first had told me. I do not want to criticize, to apportion praise or blame, really, not even the latter, but that is how it is. What is needed, particularly among anthroposophists, and I have said this on many occasions, is an absolute and unerring sense of truth. It is very difficult to continue working in all these areas unless there is a basis of truth, of genuine, unvarnished truth. With this basis of genuine truth, surely when a situation arises and one pursues the matter further, a fifth or sixth person would still present the same facts. Yet it happens that I am told about something 'dreadful' and everybody I ask tells me something different. I cannot, of course, apply the things I derive from other sources to external life; I have said this many times. It is not a question of whether I know about it, know who is right and who is wrong. The question is whether the first says the same as the sixth or seventh. What I know has nothing to do with it. As a rule I do not allow people to pull the wool over my eyes, and that is not why I ask people. The reasons are quite different. As a rule it does not interest me very much what people tell me. The point is that I hear what the first person says and then the seventh, only to find on many occasions that one person says one thing and the seventh says the opposite. It evidently follows that one of the two things cannot be true. It seems to me that this does follow, after all.

In outer physical life, which for this very reason is succumbing to decline, people have always wanted to shut their eyes to the function, the crucial significance, of untruths. Even unintentional untruths are destructive in their effects. In anthroposophically oriented spiritual science it is absolutely essential to realize that an untruth in the life of mind and spirit is the same as a devastating bomb in physical life. It is a devastating force, an instrument of destruction, in very real terms. It would certainly be possible to do important and fruitful work in the spiritual sphere again, in spite of the many new developments, providing these things are given some attention—objective attention, however, not subjective attention.

You know I do not normally go in for tirades; it is not my habit to moralize. Just for once, however, I really must discuss the facts that have become so evident at this time, because the situation is serious. We are looking at undertakings that must not fail, that will have to succeed, and there can be no question of any kind of failure; we have to say today that they shall succeed. But they must not swallow up the original anthroposophic movement, and this means that everybody must do their share to ensure that the moral foundation established in the work of many years really persists. Everybody must do their part. It is really necessary for everybody to do their part.

It saddens me deeply that I am unable to respond to most of the many requests that are made to me. I had to keep refusing to help members because time cannot be used twice, and meetings go on not only from morning till night, but even well into the night. Quite obviously I cannot use the same time to talk to individuals. The membership in the widest sense must come to its senses and remedy things such as those I have just been mentioning, that are playing such a role in everything happening here. Every single member should reflect and see that these things, here in this place, must be overcome and eradicated. Unless this is done—and these things are connected—it will not be possible to find the time to do real, fundamental spiritual work. Everything arising out of anthroposophy will succeed. Yet unless some things change the primary spiritual movement will be swallowed up. The will impulses of those who consider themselves the bearers of this spiritual movement would then lead to a new materialism, since the original spiritual movement will have been aborted. The spirit needs to be nurtured or it will die. Materialism does not arise of its own accord; you cannot create materialism, just as you cannot create a corpse. A corpse is produced when the soul leaves the organism. In the same way everything created here on a spiritual basis, out of something that has soul, will become entirely material unless there is a genuine desire to nurture the spirit. This means above all giving careful attention to the moral and ethical basis which we have been able to establish. We must ensure above all that we do not succumb to illusion, that we do not think it is enough to accept certain views just because they are easy to accept. We must look at life without flinching.

It is really very bad for people to say things like: 'The threefold order is a good thing; we must adopt it'; to feel good about this, and say: 'I am organizing something and it is very much in accord with the threefold order; how good I am! It makes me really feel good to get something organized that represents a nucleus of threefolding.' Moral self-congratulation, inner self-gratification—you may feel like this when you are doing such things, but it does not mean that you have a sense of reality. The threefold idea is true to reality because it requires genuine effort to bring it to realization. Many people's ideas are, however, so unrealistic that the idea of threefolding goes against the grain for them. The first and most essential thing is for this idea to be taken up by a sufficiently large number of people. We must have the necessary sense of reality and practical common sense.

Eight days ago I had to speak here in Stuttgart about the consequences the threefold order has for the management of landed property.[40] I said that the threefold order obviously aims to achieve a situation where social exchange, social conditions relating to landed property, are such that land cannot be bought and sold like other goods. That is entirely based on reality; to say the opposite would be unrealistic. I had to discuss the subject on a day when I actually got here late because we had been driving round the countryside all day trying to buy land. If we have a sense of reality the threefold order cannot be founded on us saying, 'I must be good since I am forming a nucleus for the threefold order.' No, we have to put aside all illusions and accept that the most important thing in working for a threefold order is to work out of actual, present conditions.

It is not a question of morally congratulating ourselves as we say that we follow a particular idea. This would make it unfruitful and abstract. It is a question of seeing the actual reality, seeing what is necessary. This is the difference between people whose approach is utopian and dogmatic and those who take a practical view. The latter will take an idea as far as it can go, but they are not unworldly people living for some private pleasure; they engage with reality. Giving oneself up to illusions serves only one's own private pleasure. This must be realized. We also have to realize that many other things tend in the same direction. I am sorry, it could not be helped: there were quite a number of other things that I could have talked about today before my departure, but I could not avoid dealing with matters which, though they were put to me more or less in passing, do have an impact on the fruitfulness of our activities. One of the main hindrances to fruitful activity is the constant need to have endless discussions on matters that should be dealt with in half an hour, because the waters are muddied by things that really should not be present. If you have sound thinking habits—and those are the habits we must acquire if spiritual science as it is presented here is to come about—and then find yourself (I am not speaking theoretically) in the midst of what is nowadays called 'business practice', the best way of defining what goes on is that people kill as much time as possible, that time is wasted. There are practical people today who boast of being busy all day long. If they did not waste so much time, their work, which let us say takes ten hours, could be easily done in one hour. Nowadays time is wasted particularly on what is called practical life. This wasting of time makes thoughts diffuse and dispersed. Entering into practical life as it is pursued today one really gets the feeling that one is in a noodle factory where thoughts that ought to be concentrated are drawn out, pulled apart like strudel or noodle dough; everything is stretched out endlessly. It is dreadful to encounter the fluffed-up thoughts cultivated in practical life. If you wanted to use thoughts like these to get a clear understanding of the world, of the things I have spoken of today by way of an introduction, you would get nowhere. All this fluffy thinking has arisen from time-wasting. Thoughts that ought to be concentrated—for that is the only way for them to be effective—simply come to nothing by being drawn out. Something which functions properly at a certain density will of course be useless once it is extended too far and grows thin and tenuous. Many of the things that play a big role in modern economics are quite useless when it comes to progressing world affairs. Our particular task would be to grow concise in our thinking, also in relation to practical matters, and not to waste time. However, time continues to be wasted these days, as long as the anthroposophic movement, which after all supports our enterprises, has not become what it ought to be: a movement based on truth in every respect, a movement where all untruth eliminates itself because we have no use for it and because it would immediately show itself to be what it is.

This is what I wanted to say to you today. It is not addressed to anyone in particular. Please do not go round again saying that I had a specific thing or person in my sights. I wanted to characterize realities in general. The world situation is serious today and the things that have been going on among us here in Stuttgart really reflect the serious situation that exists for the whole of civilization. The things that haunt us in our community here can teach us a lot about the things that haunt the world as a whole.

I do not wish to hurt anyone's feelings. Nor do I want to moralize, to preach at you. My intention has been to discuss things which kept striking me over the over the past two weeks, albeit only indirectly.

Lecture 6

STUTTGART, 25 JULY 1920

THERE has been a basic theme to everything we have been considering here recently. Our point of departure has repeatedly been the need, when any work is undertaken or any proposal made in connection with the anthroposophic movement, to pay proper regard to the gravity of the times. Everything I have said so far should support increasing numbers of our members to feel this as an underlying reality in their souls. We will continue along these lines. Today I want to refer first and foremost to something that can help us find the right inner attitude, as it were, to the anthroposophically oriented spiritual-scientific movement.

Clear evidence has now been presented for the decline of Western culture—you know of the book by Oswald Spengler.[41] How do people regard the search for knowledge within this whole, developing current of Western culture, irrespective of the degree to which they even admit this? People who imagine they have both feet firmly on the ground, considering themselves to be eminently practical people, regard the search for knowledge as something theoretical and not as a real deed accomplished by the soul. It is essential for us today to recognize that the search for knowledge is a deed accomplished by the human soul. We must come to realize that when we gain knowledge and insight this is no mere theory, no individual point of view, but an actual deed infused with will impulses, a deed within the overall context of the evolution of the earth and humankind.

To begin with let me use a more methodological approach to show what I mean by knowledge as a deed, illustrating this with a reality of the world of spirit. I have frequently spoken of two streams in the life of the human soul that tend in opposite directions. One of these is the abstract mystical stream, the other the abstract materialistic stream. The latter has developed with the evolution of science over the last three or four centuries. Basically it has entered into all areas that play a role today in the progress of human evolution. The traditional religious creeds, as they are officially represented today, scarcely play a role in the real progress of human evolution. They certainly do play a part, however, in furthering the decline of Western culture.

If Spengler's idea of the decline of the West were to be fully realized, traditional religious faiths officially represented by the Jesuits, by positive Protestantism and so on, would certainly play their part. They have no role to play, however, in progressive evolution. As I have said on a number of occasions, the materialistic stream is clearly in evidence even in people who themselves are quite unaware of this. Characteristically, and it is something we must keep in mind, even certain circles of the theosophical worldview were affected by materialism. The descriptions given of the human etheric and astral bodies in those circles, where these bodies were merely said to be more subtle forms of matter, with people imagining some kind of mist or cloud, surely were nothing more or less than materialism in spiritual guise. Spiritualism is of course materialism most heavily disguised as something spiritual, for it speaks of the spirit when in fact its aims are merely to prove the material existence of the spirit, to present it in material form. Materialism has infiltrated everything disseminated as popular literature, above all popular books and journals where people are informed what is 'true'; it is present in everything that is put about like this, irrespective of whether it comes from Catholic or Protestant sources. This materialism on the one hand relates to cultural progress. It must be taken into account and taken positively into consideration. Traditional historical elements like the religious confessions must of course attack anything that is new; they must fight intensely against anything that is new. This, however, since it tends towards decline, does not have to be given serious consideration by the contemporary mind. Materialism on the other hand produces the very things we ought to know about today, though they are of course presented in a materialistic way, in materialistic interpretation.

If we wish to share in the work that brings progress in cultural and intellectual life we must know what materialistic anatomy, materialistic physiology, materialistic biology and the sociology of the present age have shown us. We must be fully informed about these things and out of this very knowledge gain the power to transform materialistic knowledge, the materialistic way of thinking, into spiritual knowledge. It is therefore of definite value in the present time to give full consideration to what materialism has to offer. We cannot transform, say, the Catholic philosophy of the Middle Ages in the way some people imagine. This can only be transformed with the aid of Thomism, as I have shown in Dornach,[42] through which it then transforms itself. Materialism can however be metamorphosed into an inner spiritual life. Anthroposophists therefore have no reason at all to despise the things that materialism has to offer. We have to reckon with materialism. Anthroposophy cannot be developed out of airy nothings; it must be developed by people who are alive in, and part of, modern life, a life that in the first instance is a materialistic one.

The moment we wish to see materialism in the light of the true progress of humankind we must develop a particular underlying feeling, the very feeling that many people of the present age, and above all academics of the present age, do not have. This is the sense that everything immediately around us in the world we perceive with the senses, everything our eyes see, our ears hear and so on, is not real and that we should not look for reality in that direction. We must develop the feeling that it is utterly mistaken to look for atoms and molecules in the world we perceive around us and to consider them to be real, or even a valid currency of thought. Some scientists are particularly proud to say that they do not take atoms and molecules to be real but use them as thought-forms, ideal points in space. It is immaterial, however, if you assume atoms to be physical or ideal points. What matters is whether you take a living comprehension of spiritual entities as your starting point or whether you consider the idea of such living comprehension an abomination and base your ideas entirely on what may be derived from the material world. This applies also to atoms seen as the loci of forces.

As soon as you base your views on atomistic ideas you find yourself in a materialism that must lead to decline. We can only deal properly with the world we perceive through the senses if we treat it as a phenomenon, as a form of manifestation. Matter is not even present in the things we perceive through the senses. We must therefore develop the feeling—we can do this thanks to the research results recorded in anthroposophic literature—that when we use our eyes and look out at the whole starry firmament, the cloud formations, the contents of the three kingdoms (mineral, plant and animal) and also the fourth kingdom, the human kingdom, we must not look for anything material in the things that come to us through sensory perception. Matter does not underlie any of them! All we perceive are phenomena like the phenomenon of the rainbow, for example, though they may appear more solid than a rainbow. No one should consider a rainbow to have some kind of outer reality—like a solid bridge with its span of seven colours—but only a phenomenon. In the same way we should regard all the things we encounter through our senses as phenomena, however solid they may appear. A rock crystal can of course be held in the hand, whilst we cannot do this with a rainbow. Yet although the crystal impinges upon our sense of touch, it should still be called a phenomenon. We must not allow our fantasy to create some kind of physical reality, in spite of the view of nature that is generally taken today, a view that pursues the wrong path. The 'physical' phenomena we come across therefore are definitely not material phenomena, are not matter in reality. They are mere phenomena; they come and go out of another reality that we cannot comprehend unless we are able to conceive of it in the spirit. That is the feeling we must develop—not to seek matter in the outer world.

This is why those, above all, who despise outer materiality, miss the real goal of anthroposophic development: people who say: 'The things we perceive in an outer way are mere matter; we must rise above such things!' That is quite wrong. The things we perceive outside us are not material, we cannot look to them to find the world of matter. We simply do not find matter in the world that impresses itself on the senses. You will come to see this if you read what our anthroposophic literature has to say, and take it in the right spirit.

You then need to develop this feeling further. Here we come to things that people find highly uncomfortable today because they come very close to the experience we know can be had with the Guardian of the Threshold. They are uncomfortable; yet unless we enter into them we will make no progress in inner development. We have to go through inevitable discomfort if we are to get from theory to reality. The search for knowledge must be based on realities. Anyone who thinks matter can be found in the world which we call the material world—the world we perceive with the senses—is mistaken, and the error involves more than mere theory. There are people who think that simply *calling* something 'matter' is enough to make it so; this kind of word-cleverness is in vogue nowadays. And likewise if anyone thinks it is enough just to assert it is wrong to seek matter in the world we perceive with the senses, they too cannot be said to be speaking out of anthroposophically oriented spiritual science. Spiritual science does not consist simply in correcting other people's theories. Spiritual science must make the search for knowledge into a deed. It must be a search for truth based on strong will impulses; that is, it must engage with realities even at the point where it creates definitions or offers explanations. And this is where things get uncomfortable.

It is easy to say to someone that they are wrong in thinking that matter is to be found in the outside world that we perceive with the senses, and to tell them to change their views. That is just talking theory. To accept theories, to oppose theories, to correct them—all that is theoretical talk. Spiritual science cannot in all reality be satisfied with this. The essential thing is to develop our sensibilities to a point where we perceive that someone caught up in materialistic views of the material world has a thoroughly unhealthy organism. We must progress from purely logical definition to a definition that takes hold of realities, in this case the constitution of the human individual. We must become convinced that it is not merely logically wrong to say that matter is to be found in the world we perceive with the senses, but rather that thinking what our senses perceive to be physical substance is actually a path toward constitutional feeblemindedness. We must perceive that to be materialistic in that sense is a kind of illness.

We want our ideas to take hold of reality. We cannot do this whilst we continue to think in theories. Everybody supposes that they only need to receive good instruction to change their views. Spiritual science always requires us to be alive as we develop, to restore ourselves to health if we have been materialistic in the sense suggested, since this kind of erring means sickness, the road to feeblemindedness.

At this point things come very close to the insights to be gained in meeting the Guardian of the Threshold. When we encounter the Guardian of the Threshold and thus enter into worlds other than the physical world—worlds that complement the physical world with a new element—all theory comes to an end, the intellectual mists clear and reality begins, and every word becomes saturated with reality. Then we can no longer say that someone is expressing correct or incorrect views. We have to say that the views they express come from a sick or a healthy mind. Then we encounter reality. Nor can we say that someone should correct their views. Instead we must say: 'If you are on the road to sickness, to feeblemindedness, you must change course and develop a strong, healthy mind again.' You see, it is not enough to correct the so-called philosophies that veil everything in their miasmas. For anyone wishing to become a spiritual scientist it is essential to pass through a change that is a very real process, and not to be satisfied with something that is intellectual, logical or theoretical. The gravity of the present situation is such that the pathological nature of an intellectual view of the world must become vividly apparent to us.

I have attempted here to outline one particular aspect, to characterize in the light of reality the materialistic aspect of our cultural life today. The other aspect, the polar opposite of this, is the mystical approach. Mysticism is the refuge of many people who are dissatisfied with materialism. They find that materialism is not right and therefore feel they must follow a different philosophy, a different path in their search for truth than the paths followed by materialism. People then try to pursue an inner development by following an inner path, seek to find the spirit on that path. I have frequently spoken of mysticism as a spiritual stream that has the same right to exist in its one-sidedness as materialism has, providing one perceives this one-sidedness. I have spoken of mysticism as a kind of reaction against the materialism which has developed in American and European civilization over recent centuries. I have referred to this a number of times, also in the pamphlet published during the war that was sent to soldiers at the Front.[43] This mystical stream must be considered in more detail, again without any of the theorizing that is so pervasive. When it comes to mysticism, people think that by withdrawing from outer life and entering deeply into their inner life they will find the spark of which Meister Eckhart spoke.[44] They think they will come upon the revelation of the true spirit that cannot be found in the outer material world. Mystics do however tend to be real materialists. Though they take the opposite route, mystics mostly are out-and-out materialists. They start to castigate the material world as soon as it is mentioned, considering themselves superior to such things—as has often been said, they feel they are above such things. The point, however, is that we must not merely theorize but enter into reality. The point is that we must look for the reality behind such mystical endeavours. It is important to realize what comes to life in us when we become mystics, what is active in us when we become mystics. You can read about this in anthroposophic literature. We have to say that this is the very soil where physical matter is to be found. We find materiality active in us when we become mystics. Consider even the most sublime mystics—what are they bringing into play in their souls? They bring into play things that boil and bubble in their metabolism, however refined and subtle this metabolism may be. Matter as such is to be found within the human skin, and not in the outside world that impinges upon us through the senses. We come upon physical matter when we allow things kindled in the metabolism to arise within us. Look at the way Meister Eckhart spoke of God with such depth and conviction. What he actually conveys is his conscientious awareness of what was bubbling and boiling in his metabolism. It seemed to him to work towards the heart centre and there to become transformed into something that could be perceived as a spark of the divine self in the human being. That is the small flame which metabolism ignites in the heart.

The true nature of physical matter is thus found by following the path of mysticism. In the same way that the genuine fruits of Goetheanism must be raised to a higher philosophy of life, so we must clearly understand that the fruits of mysticism are something whose interpretation must be sought in material activity. We do not discover material processes in our chemical laboratories. When a chemist is at work in the laboratory, the processes taking place in the test-tube are external phenomena, just as a rainbow is an external phenomenon. That, too, is phenomenon and has no real materiality to it. We learn about real materiality when we see the bubbling and boiling of the processes that go on inside our skin, igniting in the same way a wax candle ignites to burn with a flame. That is where materiality has to be sought, and we only see mysticism in its right light when we realize that all the inner experiences mysticism provides in its one-sidedness are material effects; true materiality is to be sought there. We must not look for physical matter by analysing chemical processes. We must look for physical matter in the complex chemism and physiology accomplished within the human skin. Mysticism gives us the solution to the riddle of physical matter, but *only* gives us the solution to this riddle. We must not reinterpret the inner materiality of the human organism to the effect, for instance, that when we see a burning candle we say: 'That cannot be the fruit of something inherent in the candle. There is a tiny spirit inside that candle and this spirit produces the flame.' That would be nonsense of course. It is also nonsense to look for the reality of the spirit in the experiences of a mystic.

We have to arrive at a very definite idea, even if this is difficult. This is a threshold truth. We do not get far with what can be achieved in mysticism, for there we are dealing with phenomena that are like opiates, we are given up to our egotistical desires that allow themselves to be defined as anything but the materialistic aspects of our own inner processes. The bewildering multitude of phenomena surrounding us in the world of the senses does not allow us to go so far as to realize that in fact none of it has any materiality. Let us consider what we are actually seeing when we look at a distant planet, say, or a fixed star out there in space. What are we actually seeing? Naturally we do not see the earth's green plant cover, cloud formations, brown or grey earth and so on, such as we see around us on this earth. The stars and even the moon are too far distant for that. Everything that lives out there on those alien heavenly bodies has an inner aspect, possesses material processes that have been transformed. What we see through the telescope are the material processes active in the highest form of existence on the star or planet in question. In the same way, if that other star or planet, let us say the moon, were to look at us through a telescope, would it see our plants, animals and so on? No, the earth is too far away for that. Pointing a telescope at the earth, the moon would be examining your stomachs, hearts and so on. That is the content which shines forth into the cosmos. The human kingdom is the highest on earth and, because of this, someone looking from outside would see what goes on within human skins. And when these things which are visible to distant stars become kindled in our own inner awareness, they are the things which mystics experience.

So you see that anyone seriously devoted to the anthroposophic view will have to penetrate this second, equally uncomfortable threshold truth that mysticism teaches us what matter is on earth. We cannot know anything about even the simplest of earthly forces if we merely look at the outside world. Just open a book on physics and read about gravitation, earthly gravity. This will always also includes the comment, however, that it is impossible to know the true nature of gravity. People are in fact rather pleased with themselves when they explain that the essential nature of gravity is unknown.

How can we get to know the nature of the force that makes a piece of chalk fall when I let go of it? The force called gravity can be understood as follows. At a certain point in life, perhaps after the age of thirty, or even earlier—it depends on how kindly destiny deals with us—we can make a discovery when we observe ourselves in the light of spiritual science, rather than by the usual methods of observation. The methods of spiritual science do to some extent introduce us to genuine self-observation. Around the age of thirty-two, therefore, we can make a discovery. Observing ourselves not in the abstract way of mystics but genuinely, we shall achieve genuine self-observation by recognizing, for instance, that between age thirty-five and forty, let us say, we have changed at the organic level. Some will note that their hair has turned grey; perhaps they have grown bald. We find we have changed. Unless, however, we have gained the ability to observe ourselves we shall not inwardly experience what accompanies these changes. But if people apply to themselves something contained in my book *Knowledge of the Higher Worlds*,[45] they can have this experience. From about the age of thirty-two onwards, we can find that the body has to be carried differently, that it becomes heavier. Then we have an inward experience of gravity. This is something that has to be experienced inwardly.

None of the wishy-washy things talked about in mysticism are as important as a tangible fact like this, the inner experience of growing heavier. You cannot gain this experience if someone stands here and lets a stone fall from their hand. You do not observe the gravitational pull by watching a stone drop, for stones have no real materiality. You must observe yourself, looking not into space but into time; that is, observing a succession of events. We must progress from experience in space to experience in time Things never to be found in the world of the senses must be gained through inward experience. They are the second element belonging to reality.

Experiencing the outer world of the senses we possess merely a truth, but not scientific discernment. Experiencing inwardly in an abstract, mystical way we have merely discernment but not truth, for we are subject to an illusion concerning the basic phenomenon of inner life: of inward experience being the flickering flames of material processes. Anyone looking for materiality in the outside world is interpreting the world in an ahrimanic way. Someone else may merely look for truth in an abstract way within; they are interpreting the world in a luciferic way. Genuine spiritual science holds the balance between the two, an interweaving of truth and scientific discernment, of truth and science. We must look for truth at one pole and science at the other, and become aware that living realities become polar opposites when science informs truth and truth informs science. Then the search for knowledge becomes a real deed. Then something actually happens: we are not merely producing logical definitions or correcting views, but something is happening when human beings endeavour to gain inner experience of science, and seek truth outside them, endeavouring also to let each penetrate the other.

This has to be understood in the present age. The present age needs to understand that human beings must hold the balance between the two extremes, between the ahrimanic and the luciferic poles. People always tend to go in one direction. In the sculptural group in Dornach,[46] the luciferic element is above and the ahrimanic below. The Christ stands in the middle, holding the balance. These things can be presented as ideas, can be distilled into ideas. They then become truth and science. It is also possible to represent them in art, but then we have to forget about mere ideas and seek to find them in line, in form, in configuration. Then it becomes the Group in Dornach, for instance. The whole is, however, permeated by the spirit.

Mysticism is one-sided and so is materialism. We must know that the two have to be interwoven in a living dynamic, and that in this living activity we must seek our true inwardness. Our age wants to be one-sided and embrace materialism and this means that it is pursuing a path toward feeblemindedness. I have shown that we must not be content with theories but must know in truth and reality that materialism shows itself to be what it is—a road to feeblemindedness—as soon as we meet the Guardian of the Threshold. We must aim for a state of health, and not merely disprove things in order to arrive at something else. The opposite extreme is abstract mysticism. We should be able to develop the feeling that in reality this is the road to infantilism, childishness—a condition appropriate only for small infants. A child as yet untouched by the world, living entirely in physical materiality, in the processes of its physical organs, is exactly the type of the mystic, though the mystic will have the same experiences at a later stage than a child. They will of course feel different, those experiences, but an infant also experiences this concentration of organic activity in the heart. Sensing this concentration it will kick its legs in the air and wave its arms about and we can see how this peripheral activity is the opposite of a concentration of activity in the heart. If people remain childish all their lives, if they are too lazy to absorb the things that only materialism can give, they reject outer materiality; it means nothing to them, they see it as a lower state that must be overcome. And then they kick their legs in the air and in doing so produce their mysticism. That is the threshold truth, the unpalatable threshold truth. Everything that is abstract and mystical, inducing a feeling of self-gratification when people concern themselves with mysticism nowadays, with things that make them congratulate themselves when they appear in print, though in reality they are the equivalent, in thought, of just kicking one's legs in the air— all that is infantile. It has to be clearly understood that whereas materialism leads to feeblemindedness, abstract mysticism leads to infantilism, to childishness. True life is found when we establish the balance, the equilibrium, between materialism and mysticism.

Again it is rather difficult to do this, and things become uncomfortable. When you want to balance the scales you must not despise anything that is present in excess on one side and upsetting the balance. You must really try to put into both scales what is needed to maintain equilibrium. In the same way you should not despise anything that takes you into the sphere of matter, saying to yourself that it will cause feeblemindedness. Quite the contrary: those wishing to engage with things must step boldly into reality, saying to themselves: 'I will have to follow the path that would lead to feeblemindedness if I were one-sided in my pursuit; but I am armed against it. I am also armed against pursuing the other path one-sidedly; I retain what is necessary from childhood days but do not remain a child.' That is how the balance must be sought between materialism and mysticism. That is a true sense of life. The sense of life holds the balance between feeblemindedness and childishness. If we cannot be bothered to see these things clearly, we will not be able to engage with reality. People only grow feebleminded if they fail to recognize that it is normal for people to have to overcome feeblemindedness day by day, hour by hour. Feeblemindedness is a constant threat and we only remain human by remaining childish, i.e. inspired. Anyone holding on to childishness in the right measure is a genius. We are geniuses only to the extent to which we have held on to childishness into our thirties; but this childishness must be properly counterbalanced. Thus we have to say that we are all in danger—how shall I put it?—of becoming geniuses or remaining childish infants. It could go one way or the other.

As soon as we come close to threshold truths, our ordinary ways of expressing ourselves no longer work; things that normally are quite separate blend into each other. All words acquire a different meaning, and we might say—and it would be an amusing thing to represent this in a painting or sculpture— 'Here is the threshold of the spiritual world, with one individual on one side and one on the other; one is active in the spiritual sphere, the other in the material world, and they are yelling at each other. The one who is in the spiritual world yells: "Childishness!" The other yells across from the material world "Sheer genius!".' Just as a tree looks different when seen from different points of view, so things appear different depending on whether we look at them from the spiritual point of view or from the perspective of materialism. From the spiritual point of view the genius of someone who has retained the ways of a child, playfully forming ideas, has to be called childishness; we must see it as childishness when viewed from the spiritual perspective. Childishness is regarded in a different way from that point of view. There we know that human beings descend from the spiritual world, that they come to live in a physical body; we see that a child is still lacking in skills, is still undeveloped, but we also see the most sublime spirituality alive in that child.

It has caused considerable annoyance to some people—that numbskull Dessoir,[47] for example—that in a small work I published, *The Spiritual Guidance of Man and Humanity*,[48] I showed that the wisdom involved in giving shape and form to the brain of a child is far greater than the wisdom human individuals are able to produce in later life. Numbskulls like Dessoir cannot grasp this. In their view, only the books they write contain wisdom of the fullest scope. The thing is, however, that when we say 'childishness', from the spiritual point of view we perceive how the human spirit has descended as a ray of the divine spirit, and that it was fully developed when it did so. It entered into a human body that was still undeveloped, taking hold of it, working upon it, with the result that after just a few months the brain has become something different, and the whole body is something different in the seventh and fourteenth year of life, and so on. Childishness is not a term of abuse, therefore, for childishness is seen to be the descent of the spirit into the physical world, a first taking hold of the body, a stage where one is still a child, still in a condition in which the spirit has not yet been purged from the head. That will happen as the rest of the body develops, for this develops fastest, whilst the head contains far more spirit. That is the image we have when we speak of childishness from the spiritual point of view. The head of a child is full of spirit and—this is an unpalatable truth—as we get older the spirit diminishes, our heads become more and more petrified. A child still has a great deal of the spirit. This gradually evaporates. I may be permitted to use the term 'evaporate' in the sense that the spirit evaporates from the head down into the rest of the organism. So you see I am speaking of something most sublime when I speak of childishness as it is seen from beyond the threshold. If I speak of childishness from the earthly point of view it means that one has failed to progress. The language of the earth and that of heaven are different, alas, and it is part of the tragedy of our age that people do not even want to understand the language of heaven. Since it has become customary to preach from pulpits in the most earthly terms possible, it is no longer possible for people to understand the language of heaven.

It can easily happen, when one has something to say in a certain context—formulating it within that context, of course, and having prepared listeners before saying words that come from beyond the threshold, words to the effect that entities of the spiritual world evaporate downwards—that the following may occur. Let me present a picture to you of something that really happened. It may happen, then, that someone writes: 'Steiner says that things evaporate in a downward and not an upward direction.' Some professor of anatomy[49] gets hold of this and reads it out to an audience whom he himself has invited to bring along children's trumpets and rattles to a genuine lecture on anthroposophy. Thus someone gives a lecture on anthroposophy. Then the professor takes the stage and reads out a quote like the above, having somehow got hold of it, and his students use their trumpets and rattles to produce the kind of scientific argument that has become customary in those circles. This is something that really happened in Goettingen the other day. Have a look at the supplement to the recent issue of our *Threefold Order* journal.[50] You will find it there.

These are serious times in which we live and on Friday I want to continue in the vein in which I started today, when I characterized the true face of materialism for you on the one side and that of mysticism on the other. I will then show you what we are called on to do. We are not called today to gather in sectarian groups, but to enliven all that occurs in the world, bringing anthroposophic impulses into the cultural life of today. If we understand what the present age asks of us we cannot remain one-sided materialists or mystics, we must take the road to reality. I have tried to characterize this in the pamphlet Herr Molt took the trouble to publish for soldiers at the Front, so that they might learn something of the anthroposophic spirit. We must always keep in mind that these are serious times in which we live and that we shall only feel able to cope if we are open to the approach of something that properly speaking cannot even be expressed in old forms of speech, but requires us to find new formulations to become a truth fit for contemporary conditions. The search for knowledge must go beyond mere rumination, it must become an active deed. Then humankind will not slither into the decline of the Western world, for we shall find the upward path again. As long as materialism continues to use the symbols of childishness—those trumpets and rattles—to rebut anthroposophy, and mysticism makes use of materialism, dressing up utterly material processes as something spiritual, we shall slide at full tilt into the decline of the Western world. It is not a question of ruminating but of really doing something.

Lecture 7

STUTTGART, 30 JULY 1920

TODAY I will continue with some of the topics I discussed the last time I was here. It is particularly important, indeed vital, to stress the connection between what I have said before and what I wish to add today. I said recently that the road to spiritual science calls for recognition of two facts. One is the impossibility of imagining that matter, physical substance, can be found in the outer world of our human environment. This can be clearly understood through many different things that can be learned in spiritual science. Our eyes behold the outside world, our ears hear the outside world, and we gain a certain understanding of nature when we use the intellect to combine what we see, hear and perceive with the other senses. We then think we know something about the natural world outside us. Yet we are in error if we think and believe some form of science will help us to find physical matter and the laws pertaining to it in that outer world of nature. Materialism's mistake was not that it spoke of physical matter but that materialists thought they could find physical matter and the laws of physical matter, its infrastructure and essential nature, in the outside world. People who say, therefore, that they do not want to know about the external world because it is a material world, that they want to pursue an inner mystical path to a world of the spirit, are materialists just as much as those who simply interpret the outside world in materialistic terms. Their mystic search shows that in their view, too, physical matter is located in the outside world. The error of modern times is to look for the essential nature of matter in the outside world. To put things right essentially means that we must no longer look for the nature of matter in the outside world but be very clear in our minds that however far we extend our sensory perceptions we shall never discover the nature of matter and its infrastructure, its laws. It has to be understood that all that exists in the outside world is *maya*. It is the world of phenomena. Look as we may we shall never find anything material in that external world.

On the other hand we must grasp a second, quite different fact. It is that the nature of matter, which materialism seeks in the outside world, is found within ourselves. We shall find it particularly if we become one-sided, abstract mystics. The contents of a certain mysticism coming to our awareness—experiences we think we are having—are nothing but the flame, I would say, that is lit within us by processes involving our physical organs. If we study the mysticism of Tauler and of Meister Eckhart, we will be right in thinking that these men had a special faculty for experiencing such things, and for interpreting the physical matter within them when the flame of awareness was kindled. They found the material world through mysticism. Until we know that external observation reveals only the world of phenomena, maya, and that inward observation reveals only physical matter and its flame, we cannot get a clear, true picture of the nature of the world and the way human beings relate to this world. Physical matter is not to be found by applying science to the outside world, it must be sought within us, through mysticism. There we shall find its laws. The essential nature of gravity is not to be found with the aid of Atwood's machine.[51] Instead we can try—once we reach the age of thirty-two, or perhaps at another point in our lives—to become inwardly aware of gravity, so that we know from inner experience what it really means to experience gravity. Tangible inner experience should show us that between our thirtieth and fortieth year we grow inwardly heavier and heavier. We can gain inner experience of a property of matter that merely comes to expression in mystical experiences. I have tried to demonstrate the essential point by saying that anyone immersed in the chaos of the planet, as modern scientists are, cannot get a clear idea about these things. We see the plants, the animals, the cloud cover; we see the glittering light of the stars, we see rivers, hills and valleys and so on. Yet if someone were to observe the earth from Mars, for instance, none of these would matter. An inhabitant of the planet Mars observing the earth through some instrument or other—in a different way, those who inhabit Mars do, as we might imagine, have a kind of organization enabling them to observe the earth—would perceive nothing of the cloud formations, rivers and mountains we see, nothing of the phenomena relating to the mineral, plant and animal kingdoms. He would only perceive what goes on within the skin of the human beings living on earth. Everything else would vanish before the eye of an inhabitant of Mars. He would perceive only what goes on inwardly in the organic life of human beings, and for him that would be the material world of the earth. When we grow aware of a mystical element within us it is not what many mystics think it is but the flame brewing within us. That is the place where we can find out about the physical matter of the earth. This form of self-perception takes us into the sphere of matter and energy, a field in which the people of the Western world have arrived at exactly the opposite view over the last centuries.

This gives an indication of the extent to which we have to change our thinking if decline is to become an upward movement again. People think they are materialists or idealists or spiritualists because they follow a particular philosophy. That is not the case. We are far from being spiritual when we say we contemplate the inner and not the outer life. It could indeed happen that, precisely by focusing on their inner life, someone comes to observe matter as it becomes flame inside them. To find the right path it will be necessary to grasp what I mean, and to do so with the right inner attitude. The outer world as we perceive it with the senses offers only phenomena; it does not reveal the root and origin of the phenomena. Their root and origin lies within our own skins. Anything we see externally should be regarded in the same way as we regard a rainbow. Anyone who believes a rainbow to be more than merely a phenomenon, thinking it to be something material spanning the heavens, is mistaken. In the same way we are in error if, due to the fact that our sense of touch is also involved when we perceive the world around us, we believe we are surrounded by material things and not mere phenomena. The only difference from a rainbow is that other senses are also involved.

Materiality cannot be found there, however, just as it does not exist in a rainbow. Everything outside us is phenomenon. The root of phenomena therefore lies inside us. Within our human skin the processes take place that bear earthly occurrences onward from one era to another.

It may seem highly improbable and paradoxical to modern minds but it is nevertheless true that the phenomena which surround us today, and the laws apparent in these phenomena, are the consequence not of material events but of what went on inside the bodies of Egyptians, of Chaldeans and others three thousand years before the Mystery of Golgotha. Those inner events have become outer ones. The outside world of those times has vanished, disappeared. Human bodies hold the germ for a future that may be reckoned in thousands of years. It is possible to see this by considering the natural phenomena of today, drawing a conclusion that may be bold but nevertheless revealing. People talk about the properties of the element radium. To someone able to perceive the reality of the spirit this sometimes sounds like children talking about something adult minds have long since come to understand through different facts. Modern physicists know that the radium which existed on the earth's surface up to AD 140 has since disappeared and is no longer radium. The radium that exists today has only formed since AD 140, as physicists nowadays teach. These things present themselves to human minds today to compel them, as it were, finally to relinquish the erroneous ways of thinking they had to pursue for centuries to gain human freedom.

All this requires us to consider the things anthroposophically oriented spiritual science presents in a totally different way from how we usually regard things. We must abandon mere theory and consider the reality: to progress at all levels from abstract intellectual knowledge to knowledge as active deed, to doing things, really doing something in relation to the world. As I have said before—but it is essential to make this point with real forcefulness—people think that some are materialists nowadays and others are spiritual A spiritual person will think others are materialists and should be opposed because they are mistaken to think the soul is the product of physical matter. 'What materialists say is wrong,' says the mystic, 'and we refute their arguments.' That is not the point, however. It is not a question of logic, of theories. Yet people think spiritual science is all theory.

Anthroposophically oriented spiritual science is always founded on reality; and sometimes also of course seeks reality where it is to be found, in the true realm of the spirit. People who look to the outside world and seek to find matter everywhere by the methods now used in molecular and atomic theory—it makes no difference if they see matter as points of energy or as tiny building blocks—are not merely making a logical error that can be refuted. True spiritual science has nothing to do with purely theoretical concepts. It is concerned with reality. Anyone looking for more than phenomena in the outside world is not only making a logical error but is heading for organic illness that will affect their whole being. We should not say that pursuing this path is an error of logic. We should say that anyone searching for truth in that direction is liable to organic pathology, is heading toward feeblemindedness. Anthroposophically oriented spiritual science often has to transform theoretical views into views that relate to reality. The search for clarity of ideas and concepts has nothing to do with merely agreeing or disagreeing with the views of others; it has to do with sickness and health, very real things in our lives. It therefore has to be said that a seeker who looks to phenomena for more than mere phenomena, for physical matter, is on the road to feeblemindedness, to organic illness. These are things located very much within the sphere of reality.

In the same way we cannot simply refute those who seek abstract spirituality within themselves. Someone looking for the spirit by pursuing a narrow, inner mysticism, and failing to realize that the fabric of this mysticism, when properly understood, reveals an underlying materiality, is on the way to developing organic illness, a form of infantilism: a designation appropriate when one perceives it from beyond the threshold. At the threshold between the physical and the non-physical world, where the Guardian of the Threshold stands, the quality we call inspiration or genius on this side of the threshold may justifiably be called childishness from the perspective of the other side. Coming to expression in the wrong way in the sensory world, it persists as childishness throughout life. Genius on the other hand means that a certain fount of childlike quality persists throughout life. Genius is achieved when we are able to retain into ripe old age a quality of soul that normally belongs to childhood. This is seen in its true form from beyond the threshold. If, however, this childlike soul quality persists one-sidedly into subsequent life stages, then this element, which is genius in its rightful place, becomes childishness instead. Once again we see that merely logical ideas must be replaced by ideas that relate to reality as soon as we enter the sphere of spiritual science. They must be replaced with concepts that do more than simply change our views but also alter our inner, organic nature.

Anthroposophically oriented spiritual science is thus a very serious matter. Insufficient heed is paid to this when people approach the work of spiritual science with their ordinary mental attitudes. They want to agree or disagree the way they usually do in external life; they want to continue in their habitual ways when approaching spiritual science. Anthroposophically oriented spiritual science can, however, only be taught by speaking in terms of the world beyond. There words have entirely different meanings. Gravity, which exerts a downward pull here on earth, exerts an upward pull in that world. In the spiritual world, what draws us down must be spoken of in a way that makes it the exact opposite. It is not surprising, then, that anyone taking spiritual science seriously is, to begin with, completely misunderstood by people who want to proceed in the customary way—a way that was inevitable in the age of materialism—when they approach spiritual science. The inevitable result is that things such as those I dared put to you yesterday are misunderstood.

Someone presenting their own views in opposition to Oswald Spengler would simply refute him. As a spiritual scientist one is obliged not to refute Spengler's view in the usual way but must adopt perspectives rather than pursue a linear train of argument. We have to say that Oswald Spengler speaks from a different point of view, one offering no prospects for the immediate future. We do justice to such phenomena if we do not simply refute them but show the genius they contain, speaking with inner empathy about the things that one wishes, nevertheless, to see overcome. Spiritual science has much more to do with the way in which we deal with these things than with bald statements, with the kind of mystical platitude which the person who produces it believes, even, to be a particularly inspired truth. We have to consider these things, for we are moving into an age where we have to get beyond the mere contents of intellectual life. This is something I want to stress over and over again: we must get beyond the mere content of intellectual life.

Regarding the content alone, even a fool would find it relatively easy to refute Oswald Spengler's ideas. That is by no means difficult, but it is not what matters. What matters is to discern the actual reality living in Spengler's work and show how it can be overcome in a real and concrete way. In future, in characterizing a person, it will be increasingly important to consider what they are actually saying rather than whether what they say is pleasing or displeasing to oneself. We should not consider our own sympathy or antipathy to certain contents, but whether there is a spiritual quality to them. It is more important for the overall outcome of world evolution for someone to be an inspired materialist, a genius in presenting materialism, for that calls for a brilliant mind, whilst it often needs very little intelligence to present platitudinous mysticism. A platitudinous mystic may on occasion do more to make the world materialistic than an inspired materialist. It is the quality of mind that matters. Recognition of this fact will count for much more in future than actual content. This is something we have to learn. We must not seek for the spirit as though it were a system of logic; we must look for its reality. Let me ask you this: Could you not recognize more of the spirit alive in an inspired materialist than in a spiritualist full of platitudes? These are the things anthroposophically oriented spiritual science must come to discern clearly. It is the reality of the spirit that matters, not abstract statements made by a particular person. People fail to realize how important it is to consider realities and not theories!

Some of the things we meet in ordinary life today simply have to be considered from the point of view of spiritual science if we are to get them clear in our minds. Consider the political parties which have developed in public life today. You know that the most miserable, sterile clichés are to be found in party politics. Yet to some extent we are all part of this, willy-nilly, unless we want to withdraw completely from public life or perhaps cannot vote because we are stateless and have not been given the right to vote anywhere. Everybody who has the right to vote is required to support one tendency or another, i.e. to conform to party lines. Parties are a fact of life. They go back to better times, to the English see-saw system when there was the Conservative Party on one side and the Liberal Party on the other. It may be said that all the parties that now exist are different combinations of those two shades. Sometimes the more left-wing, liberal element acquires a little of the colours of right-wing conservatism, or conservatism is coloured by left-leaning liberalism, as in the case of the Social Democrats; or conservatism turns radical, as we have seen in the present time. All in all it can be said that the conservative-liberal see-saw is the pattern on which all our parties are based. That is the picture one gets when looking at this in an outward way. The most dreadful things happen in these party-political organizations—everybody would admit this. The thing exists, however, and the question is why it exists. What does it really represent? What in fact are parties?

Everything that presents itself in the physical world is an image of the non-physical world. What is it that exists in the non-physical world that gives rise, as a reflection of it, to parties in the physical world? The matter can only be properly understood if we grasp the conditions which apply when we cross the threshold to the spiritual world. There we arrive at something very different, at the real nature of things. Here in the physical world we are idealists, sceptics, realists, spiritualists or any other kind of -ists. We are something that can be summed up in a manifesto, as a political or sociological system. In short, we are something-ists. We adopt an abstract notion, for parties always base themselves on manifestos, systems and the like, i.e. abstract notions. As soon as we cross the threshold to the spiritual world we are no longer dealing in mere logic and abstract notions, but with realities. It is merely that no heed is paid to this. You cannot give your allegiance to a party programme once you have passed the Guardian of the Threshold; you can only hold to the intrinsic spirit of things, for there everything has to do with intrinsic and essential spirit. You can merely hold to a spirit of the higher hierarchies and say: That is whom I follow, the one I unite with. Let others present their affairs in their own way, I side with this one, I affiliate myself with him. 'Siding' with one or another tendency then becomes truly significant; it is no longer merely abstract. As human beings we may say that as soon as we look beyond the threshold we find three essential spirits: the Christ, Ahriman and Lucifer. It is of course possible to prepare oneself carefully to gain comprehension of the spiritual world and then to say: I choose Christ's party, or Ahriman's or Lucifer's party. It is, however, also possible to obscure the issue, being badly prepared, and choose Ahriman but call him Christ. We follow a spiritual entity, however—everything becomes actual being and reality beyond the threshold! We are always dealing with realities there, not with any kind of programme or system.

This relationship of the human being to the non-physical world is of a very significant nature. We cannot yet put this in final and definite terms since this would be too provocative, but very few people on earth are aware that it is, basically, an illusion to follow party lines, to accept the abstract notions of parties. There is no reality to it, and when we begin to pursue realities we must in fact pursue something that lies in the spiritual world beyond the threshold. There is, however, *one* party that can certainly be said to be well aware of this secret and indeed to be acting upon it. This was said during public lectures given at Karlsruhe in 1911[52] and has brought me the hatred of the party in question. These are the Jesuits. They know very well that to follow a party programme—forgive me for using a term commonly used in Germany—is nonsense. One follows a spiritual entity in the non-physical world! That is why their exercises start with the Jesuit having to visualize the spirit whom he is to follow in the military Company or Society of Jesus. When I say that we cannot yet assert this definitively, it is because I wish to refrain from defining the nature of what is called 'Jesus' there. The point is to show that Jesuitism forms a party that follows a spiritual entity and that Jesuits are very well aware that it is nonsense to follow some party or other that goes no further than a programme pursued in the physical world. The effectiveness of the Society of Jesus is due to the fact that it trains its followers to be the soldiers of a spiritual entity. They do not say 'This is right' and 'This is wrong'. They say: 'It is part of the mission of the spiritual entity I follow; I shall defend it. I shall oppose anything that is not part of the mission of the spiritual entity I follow, even if it is logically defensible.' It is in fact just as possible to logically defend the concerns of Lucifer and Ahriman as to defend the concerns of Christ. There are precisely three logical defences and they are all equally valid.

We therefore have the strange phenomenon that, in battling against anthroposophy, the Jesuits are of course aware that anthroposophy pursues a spiritual tendency that is wholly defensible. They know full well that logical argument is no effective opposition, for it merely means playing with logic. They know that they are facing an adversary in this battle of minds and they will use all available means. It is therefore pointless to join battle by refuting the refutations of the Jesuits. They know exactly what objections we can raise; the fact that they know them and consider them to be fair makes no difference, however, for they follow another spirit than the one anthroposophy must now follow for the weal of humankind. In the realm of the spirit it is reality that counts. What counts is that one really gains clear understanding of spiritual paths, drawing on our whole being—not the human dwarf who tends to be the end product of the kind of educational establishments we have today—to arrive at such understanding. This is something we certainly can do nowadays by using our healthy common sense. The political parties that exist in the physical domain are therefore caricatures of something that rightfully exists in the spiritual world. That is what is so difficult about it. Things appearing in the physical world may be a reflection of something of genuine significance in the spiritual world. In the physical world this becomes pernicious and abominable, because every world has its own laws, and today we face the growing necessity of working our way up toward the spiritual world again. The first stage consists of caricatures of spiritual life appearing in physical life; of people setting up party banners and following party idols when in fact they should be giving their allegiance to spiritual beings.

Truth and reality in the non-physical world turn to lies and illusion here in the physical world. You see, I am not using empty words when I tell you that what matters is to transform purely theoretical things into reality whenever we wish to speak of the truths that exist beyond the threshold.

Mere refutation of materialism will not achieve anything, because the situation for the human being is that in our whole nature we are really spirit and soul. This element of spirit and soul exists even before we are conceived, before we are born. It has evolved out of our previous earthly incarnation; it has passed through the spiritual world. It now assumes flesh, creating a physical image of itself consisting of nervous system, skeletal system, blood system. So we now have two things: the human being in soul and spirit and the human being of flesh and bone that is its image. When we think our usual abstract thoughts, what is it that thinks in us? Not the human being of soul and spirit. Particularly when we think abstract thoughts, above all using earthly logic, the physical brain is thinking in us. It is important to know that when materialists say that the brain does the thinking they are quite correct as far as abstract thoughts are concerned. The physical brain is an image of the spiritual brain, and this image creates an image, abstract thinking being merely an image. It may thus be said that when it comes to abstract ideas the physical brain does indeed do the thinking.

This is simply a special instance of what I have said before. Materialism has merely discovered that the brain is thinking the thoughts that from the mid-fifteenth century onwards have become standard in Western civilization. The materialism presented by Moleschott, Buechner and Vogt[28] cannot simply be refuted by saying it is wrong. It is quite appropriate for human beings who, from the mid-fifteenth century onwards, have turned more and more to bald materialism. Human beings of the Western world are in the process of becoming beings that think only with the physical brain. The prophets of such physical brain-thinking, Moleschott and Buechner, merely stated what Western humankind was becoming. They were wrong only in so far as they applied this to humankind as a whole. What they said applies only to people living after the mid-fifteenth century, and in their case it does apply. People have got used to thinking only with their brains; it is the common way of thinking nowadays. Everything to be found in our ordinary literature, in the whole of modern science, is material thinking, is that kind of thinking. The materialists are quite right, and we could say that Buechner and Vogt would have wrongly described their colleagues if they had suggested they thought with the spirit. That is not the case; they think merely with their brains. This cannot be disputed. It has to be recognized that the road to materiality is not merely a false philosophy but something with very real effect. That is also why, when something like anthroposophically oriented spiritual science appears on the scene, such people will say: 'These are thoughts beyond comprehension; they cannot be grasped.' Well, they want to think with their brains: the thoughts of spiritual science are, however, thought with a soul and spirit element that has sundered itself from the brain. People must make efforts to wrest their soul and spirit away from the brain with the help of thoughts that have been produced in this way; they must think those thoughts through. People must make an effort to think these thoughts through, to use the opportunity that still exists of wresting the element of soul and spirit from the physical aspect of the brain. This element is in the process of becoming chained to the physical brain. People must wrest themselves free. It is not a question therefore of right views and wrong views but of a developing process. The thoughts of anthroposophically oriented spiritual science are given to the world in the hope that people who still possess a residual capacity to wrest themselves free in this way, will actually make use of it; will try to understand thoughts that are independent of the physical body, so that their souls may grow free of the body. It is therefore a question of having the will to understand anthroposophy; anthroposophy is intended to wrest the element of spirit and soul free of the physical body. Our mission therefore is not merely to refute views that are wrong; we must face the fact that a great many people want to embrace such an outlook entirely, want to be nothing but matter, and want to draw their thinking, will and feeling from matter. We seek to give anthroposophically oriented spiritual science to the world as something real, so that spirit and soul may be wrested from matter. The aim is to prevent the possibility of people losing their spirit and soul, for they now run the risk of succumbing completely to the ahrimanic sphere. People are at risk of losing soul and spirit, of investing themselves in matter and thus losing themselves as human beings, when, as I have described, the material world vanishes into nothingness.

It is not a question therefore of replacing the old with the new, but of becoming active in the search for truth. This saves the soul from slipping into mere materiality; it saves the spirit and soul element from sailing full-tilt into the ahrimanic sphere, where the I would be lost. It is not a question therefore of refuting materialism, but of saving humankind from materialism becoming true. Materialism is in the process of developing into something that is true rather than false. When people say that materialism is wrong they are not talking about what really matters. No, we have to say that materialism is becoming increasingly true; in our present culture it is becoming an outlook that is increasingly apt and correct. We may well find that by the beginning of the third millennium humankind will have developed in such a way that materialism has become the true and correct view. It is not a question of refuting materialism, for it is in the process of becoming true. It is a question of making it untrue precisely because it is on the way to becoming reality, no longer merely a theoretical error.

Certain people seek to ignore these things, wishing to make it as easy as possible for others by telling them to recognize how wrong materialism is, inviting them to turn to an abstract mysticism that will offer them everything they need. We could take up such abstract mysticism, but that would encourage materialism to become real and not mere theory. We do not need to overcome materialism because it is wrong, by using words that remain theory; we need to overcome it because it is ever more correct and we must fight against it becoming reality. This puts another face on things, and this is also where we find ourselves in the reality of the spiritual world—not with theories, but with a living approach to the truth that in the cosmic scheme of things is an active deed. People find it unpalatable to have to listen to such things, yet that is the light in which everything should be regarded, even specific events. Believe me, the old methods of combat are finished with; everything that was customary and habitual in the past is over and done with. We must consider things in the light of the spirit.

What is conservatism? What is liberalism? Here on earth they are caricatures of the spiritual world. Conservatives are followers of Ahriman, liberals of Lucifer. Having passed the Guardian of the Threshold one can see how the whole of conservatism is following Ahriman and the whole of liberalism follows Lucifer. That may seem peculiar to the sophisticated people of today. And it is because this seems so peculiar that anthroposophically oriented spiritual science is so difficult to comprehend. We shall never understand spiritual science by merely thinking it; we shall only come to understand it if every one of its concepts makes us suffer and rejoice, when we feel lifted up and cast down, when we feel despair at a word, or think a word will redeem us, when we see destiny at work in what normally appears as shadowy theory as vividly as we see it at work in things that are done in the outer world, when what anthroposophically oriented spiritual science has to say goes beyond being mere words and becomes reality. When the inner impulse alive in this spiritual science is understood and felt, it will be rightly seen why things that for a time were propounded as mere theory, because people first had to come to know about them, must now become reality, why we have to be serious about the reality that lives in the words of anthroposophically oriented spiritual science. It will become evident in our era that we need to make the substantial essence of these words become reality.

It is still the case that the real intentions underlying the Waldorf School are not illumined by reality, that they are seen too little in the sense I have tried to characterize. Believe me, what I say is not intended as some kind of propaganda, as an appeal to your emotions. Things have been said that had to be said now because humankind must know them. That is why I have said the things I have been saying. I merely wish the opportunity would arise to express these things to a sufficiently large number of people, so that they develop an inner intention to take words as realities, not merely listen in the belief that one is offering theories.

This is what I wanted to present to you on these two occasions. External events must come to echo the inner impulses and not the external, expounded contents of spiritual science. Combatants such as the Jesuits know very well what many followers of anthroposophy still fail to realize: that anthroposophically oriented spiritual science is a reality. Since they have come to realize this—they have done so for some time now, from about 1906 or 1907—they oppose it with increasing vigour. Many anthroposophists have no idea of the methods that are used, the sheer ingenuity, because there is a refusal to be really clear in one's mind of the seriousness of the situation. Words will only evoke a little of what one so greatly wishes people would take to heart; I have tried to present a little of this to you on these two occasions. If we reflect on what has been said, if we progress from reflection to feeling, letting it become integral to our whole being, then it will not become abstract mysticism or modern science. It will become the essential inner nature of our being, it will be the power that releases spirit and soul again from physical matter, it will overcome a materialism that is not wrong but, unfortunately, right.

Lecture 8

STUTTGART 21 SEPTEMBER 1920

As you are well aware, it is often said today that spiritual science cannot have anything to do with real knowledge, with genuine perception, and that it can only be a matter of faith, a subjective way of believing things to be true. This kind of attitude then leads to a distinction being made between knowledge and belief, as is the general custom. A frequent objection raised against anthroposophically oriented spiritual science is that a kind of subjective knowledge that really can only be a matter of belief—perhaps one should not even call it knowledge but merely the subjective belief that something is true—is to be elevated, wrongly promoted to the level of certain and exact knowledge, to the level of a genuine science.

This distinction that is made between science and belief is quite a recent development. The view is that science should only concern itself with things perceptible to the senses, or at most with things that can be empirically established and explored, and that definite knowledge can only be drawn from such foundations. Belief is seen as going beyond the physical realm and it is said that one should never assume that anything that is the subject of belief can be rendered certain knowledge. Thus we have science on one side, a science limited to the physical world, and a supersensible, non-physical world on the other that may be accepted by anyone who finds it acceptable but cannot be known with certainty and must remain a matter of subjective faith.

Anyone who takes life seriously really ought to feel that the supposed distinction made by so many people between knowledge and belief poses a riddle which needs solving. Fundamentally speaking, however, only initiation science can genuinely show why efforts are made today—and have already been made for a long time, for centuries—to impress upon humankind the difference between knowledge of the finite, transitory realm of the senses and belief in something infinite, everlasting, supersensible. You know that everything presented here from the point of view of anthroposophically oriented spiritual science is thoroughly scientific in spirit and makes claim to be regarded as fully equal to the science relating to the physical world. It represents knowledge, perception, of the supersensible. Initiation knowledge has to look far back into human evolution, however, if it is to help us understand why in the present age humankind has been taught that there is such a difference between knowledge and faith.

Going back a long way in human evolution we come to a time when people had a primordial knowledge—we have discussed this a number of times—that was inherited from the gods, as it were. Things such as proof, evidence to demonstrate the truth of something, were not known then. Knowledge came to people at that time when a power arose in their hearts and minds that was not the power of empty, abstract thinking, or something similar, but a power filled with divine-light substance, divine-life substance, that felt itself to be in communion with divine worlds. Human beings knew that they were connected with divine spheres; they felt this and perceived it as we perceive colours and sounds outside us. There was no need for proof, for people had direct and immediate perception. They knew nothing of proof, nor of logical demonstration. All they knew was that as human beings they were filled with what the gods instilled into them. This certainly was 'knowledge' in the earliest stages of human evolution, and it had to do with perception of the divine origin of human beings. Knowing themselves to be united with the gods, and being given the power by their initiates to look upward to this union with the gods, people were also aware of the divine origin of man. They were aware that humankind had descended to earth from the world where it had existed as soul and spirit. The divine and spiritual origin of humankind was taken as a matter of course when this primordial knowledge existed all over the globe in the early times of human evolution.

This primordial knowledge sought to develop further, however. If it had remained as it was, people would, we might say, have continued to be filled with the divine spirit for ever, but they could not have achieved freedom, the ability to make free decisions. As soon as their arms moved, they would have had to say: 'A god within me is moving my arms.' When walking, they would have had to say: 'A god within me is moving my feet.' Those early human beings felt certainty of this kind. They felt, as it were, that a divine spirit was present within their skin. That is also the origin of the idea of the human body being a temple. In early times a human being was indeed like the earthly home of a god who descended to earth to take up his abode among human beings. Human beings had to become independent, however. As a result this primordial divine knowledge gradually faded and the divine heritage grew ever less. To achieve freedom, human beings had to develop knowledge, perception, thinking, feeling and will activity through their own resources. In a way the gods abandoned them, but it was for their own good, if I may put it like this. Divine knowledge withdrew so that human knowledge might develop. In later times the whole path whereby divine knowledge that had once existed all over the globe became earthly and human knowledge, had to be tended and governed by the mystery centres. It was the task of initiates to regulate the way humankind was to be schooled, let us say, so that human beings could find the right way of growing out of that ancient divine knowledge and into earthly and human knowledge.

At a time when much of the original divine knowledge had faded and the mysteries had assumed the task of guiding human beings, instructing them so that the right transition could be made from primordial wisdom to human knowledge, and ultimately freedom, it happened that a certain number of people from various regions of the earth came together to try to impede these intentions of guiding humankind in the right way, intentions originating in the mystery centres. We can say that fraternities were formed with the mission to counter the proper course of progress. We really have to draw upon spiritual science if we are to study the activity of a widespread fraternity of human beings in post-primeval times. History does not go that far back and there are no documents to bear outer witness to that time. A fraternity developed as I describe, and adopted mystery knowledge in a particular way, still using the methods that had been employed in the mysteries to maintain contact with the divine sources of knowledge. But by that time the mystery centres where honest work was done had long since been focused on guiding the transformation of ancient, divine knowledge into human and earthly knowledge. Thus there was a time in earthly history when the rightful representatives of mystery knowledge were fully involved in guiding this transition, one that was sound and healthy for that time. Mingled with this was an element arising through efforts made by a well-organized fraternity to restore to humankind an antiquated, primal divine knowledge at a time when it was out of date, when the murmur of ancient divine knowledge was no longer supposed to reach human ears. At a time when people had grown beyond the state of possessing divine knowledge, a group existed that still wanted the ancient knowledge to continue to be widely accessible.

Why did the members of this fraternity in post-primeval times desire such a thing? They wanted to strike at the root, as it were, of the knowledge then evolving. They did not want humanity to achieve freedom. Efforts were indeed made in post-primeval times to prevent humankind from developing the faculties that would lead to freedom, and for that purpose the aim was to strike at the root of earthly and physical knowledge. These people, who may be called the 'enemies' of human evolution in post-primeval times, made the distinction between human knowledge and divine knowledge, a divine knowledge that was no longer legitimate at the time. To deluge human beings with divine knowledge, a stage which they had outgrown by that time, meant to induce a dreamy, visionary state of consciousness. Vast masses of people lived in that kind of fanciful, visionary state in post-primeval times. Their inclinations to develop human knowledge were stifled. The reason why human knowledge came to be so deficient in many respects as time went on—I have given many examples of this—and why defects have even crept into the development of speech and language, was that a form of divine knowledge was presented to people in a way that appealed to their vanity.

Let us examine the influences at work in those who sought to befog the minds of the masses and strike at the root of the new knowledge that was evolving and also at the root of a language that arose from the depths of human nature. It has to be said that the individuals concerned were completely under the sway of luciferic powers. Luciferic powers were alive in them, luciferic powers that did not want human thinking, feeling and will to descend as far as the earth, as it were. Human beings were meant to grow ever more material, but these individuals wanted to keep them spiritual, to stop them from achieving their mission on earth. The individuals concerned were the spiritualists of post-primeval times. They were against human progress which at that time involved human beings immersing their souls and spirits ever more fully in physical bodies. The individuals of whom I am speaking wanted to prevent this, however. Considering this in modern terms—for it is difficult to give an accurate idea of the state human beings had reached during the post-primeval period—we might say that more than a little of a certain unconscious untruthfulness was apparent in those individuals. The impulse to descend into the material world, to make it part of oneself, had of course been imparted by the mysteries. The Lucifer-possessed individuals of post-primeval times certainly could not deny this. They therefore did not call themselves 'spiritualists' but actually 'protagonists of the material world'—to put it in present-day terms; these words would have to be translated into the terms in which people thought in primeval times. They told people: 'You will come to materiality if you follow us, if you make use of the power we provide in the later form of divine knowledge, if you use it to strengthen your soul and spirit. You will then find yourselves the conquerors of all that this earth holds for you; you will conquer the earth quickly and easily when you have a share in the power of the gods.' Thus the Lucifer-inspired leaders of certain parts of humanity gave themselves the honourable title of 'fighters for the material world'.

These individuals created a certain schism between human evolution as it was intended and the wrong notions which they presented to humanity, notions that the ideal was for people to *conquer* materiality rather than gradually finding their way into it and dwelling within it. Their efforts were directed toward taking hold of certain divine powers through development of supersensible knowledge at the wrong time, and using this knowledge to conquer the material, sense-perceptible world.

Today we have the reverse picture of what existed in those primeval times. Certain religious confessions have started to oppose the course science now ought to take. The roots of science have in a sense been injured. The result is that science and language bear certain defects through the course of earth evolution. Science has nevertheless come about, for sufficient numbers of people were subject to the influence of the true mysteries, and these held out against corrupted initiation knowledge and the individuals whose real aim was to strike at the root of knowledge and eradicate it. Science has come about. It has taken the road I have often characterized in detail. It reached the level it did by the middle of the fifteenth century, when the fifth post-Atlantean epoch began, and it has continued to the present time. Modern initiation knowledge shows, however, that science has now reached a further turning point. Today it is ripe to enter into human freedom, as it were. Essentially modern science still considers only physical things to be valid and exact; it is only prepared to consider things that are perceptible to the senses or may be established on the basis of experiments. As I have often said, this science is now ripe to develop toward the faculty of Imagination, to grasp the inspired, the intuitive world; to find ways to experience, to grasp the spirit. The vocation of this science is to grow, and in growing to assume the form of spiritual vision. It is ripe for this today.

For this rightful progress of science it will however be necessary for humanity to develop the same conscientious approach used so triumphantly to investigate the outer fields of botany, physics, chemistry and so on, but now apply them to inner research. We must seek to transform the attitude and approach used in external science into a way of grasping the supersensible world in a living way. I have pointed the way in my *Knowledge of Higher Worlds*, in *Occult Science*[53] and other books of this kind. We must be clear that the aims we should harbour in the depths of the soul, the only viable aims for anthroposophically oriented spiritual science, differ from Jesuitism, which is more or less its polar opposite. The difference is that Jesuitism in particular wants to keep science, knowledge as such, at the level of pure experimentation and observation. Take a look—but a careful look—at the scientific literature from Jesuit sources. The approach, the way of thinking, is as materialistic as it can be. It aims to confine knowledge entirely within the world of the senses, and strictly divide the knowledge that can only be obtained by empirical observation based on the physical senses from anything that is a matter of belief or revelation. The reasoning is that no bridge should ever be built between outer knowledge or science and anything to do with faith. Anthroposophically oriented spiritual science on the other hand aims to do just that: to find the way from a science of the physical, sense-perceptible world to a science of the spirit. This science of the spirit would, however, apply the same stringent standards as the outer science of the sense-perceptible world.

The picture, then, is as follows: the science of the physical, sense-perceptible world is the root from which supersensible knowledge is to evolve, developing from the same impulses that govern botany, physics, chemistry and so on, except that they will be applied in a different field. In certain quarters it was foreseen that this was to come. It was, however, in the interests of these people to prevent it happening, and they therefore introduced something into human evolution that now presents itself as the sharp contrast I spoke of earlier: the distinction between ancient ways of knowing that in the regular course of events developed into human knowledge, human science, as opposed to a divine knowledge used to numb human minds. The sharp distinction between knowledge and belief was presented to human minds and the true aim turned into its opposite. Knowledge of the sense-perceptible world was to be firmly retained and given great emphasis. It simply has to be admitted that Jesuit literature on materialistic science is extraordinarily brilliant in the clarity of its reasoning, its sheer readability. The Jesuit literature on the material world is much more brilliantly written than the works of many others writers on the subject today. Father Erich Wasmann's[54] work on ants, for example, is excellent; you will gain more from reading it than from the pedantic, uninspired writings of other scientists. Many more examples could be given. The [work of the] Jesuits would be excellent if they confined themselves to the material world; it is a deliberate aim [of the Jesuits] to use their description of the material world to encourage people to associate knowledge with the materialistic aspect of the physical world only. The intention is to pretend to human minds that the methods used to gain knowledge cannot be used to investigate the supersensible world. In ancient times, Lucifer-possessed individuals suggested that human beings would gain mastery of the world if they made use of ancient divine knowledge, yet evolution had already gone beyond this point. Now we have the legacy of such proponents of post-primeval times in those who pretend to the world that knowledge cannot be extended into the supersensible sphere, that it cannot exceed the scope of the sense-perceptible world. In those early times the intention had been to numb and intoxicate people with supersensible knowledge. Now human beings of the same ilk want to use all possible means to push humanity into the physical world; they want human beings to be stuck in that world and grasp the supersensible world only with the nebulous impulse of faith. In post-primeval times the aim had been to inundate humankind with an excess of supersensible knowledge. Today these adversarial intentions seek to allow human beings less than the right degree of knowledge in this sphere. Past intent was to provide supersensible knowledge that was no longer appropriate. Present intent is to let people have only sense-bound knowledge, making the supersensible world an area where every individual may hold whatever views he or she likes.

What would be the outcome if the group of people to whom we are referring were to achieve some kind of victory? These are the people who deliberately make a sharp distinction between knowledge and belief. There are of course large numbers of easily led people who come across the diatribe on the 'clear distinction between faith and knowledge' and repeat it; they merely repeat it. What is all this about? The aim is to achieve the opposite of what those individuals achieved in post-primeval times. In ancient days the intention was to prevent humanity from descending completely and taking up its mission on earth. Today the intention is to keep people tied to that mission on earth to prevent their further, earth-sustained development. The very people who are now supporting materialism call themselves 'spiritual', or priests of some faith or other, representatives of the supersensible world. In those ancient times the people offering a life in the spirit that was no longer justifiable called themselves materialists. They did so in the way I have characterized. Today a large number of people who really wish to keep humanity bound to the material world call themselves representatives of the spiritual world. The most powerful source of materialism today does not lie in the ideas presented by Buechner, Moleschott or Vogt. The most powerful source is Rome and anything that is in any way connected with this centre of materialism. They achieved their aims not by saying: 'I want to encourage materialism', but by keeping people bound to materialism. This is done by letting them develop faith merely as a nebulous impulse towards supersensible spheres and making sure that no impulse enters into humanity that could lead to comprehension of the supersensible sphere. The idea that Rome might lead the way in conquering the supersensible sphere for humanity is the historical untruth of the present age. This must be clearly understood. It must also be understood that Protestantism as it has evolved out of Roman Catholicism in recent times contains much that is of Roman Catholic origin. The desire to keep supersensible knowledge nebulous by making it a matter of faith, so that people cannot comprehend the supersensible world, has strongly persisted in the Protestant Church. Quite apart from this, we can, by reading the signs of the times, see clearly that Rome will overcome the Protestant element, and will continue to make great efforts in the direction I have described.

So you see that if one wishes to achieve something in the world that counteracts the due and normal progress of humanity, one calls oneself by the opposite name, as it were. Humanity must learn to get beyond a trust in mere names, and is indeed in the process of doing so. Humanity must go to deeper sources than words and phrases. Basically this is already beginning to happen. Imagine that someone calls and you are presented with a visiting card bearing the name 'Ernest Miller'. You would not of course expect to see someone come through the door whose clothes are covered in flour. Nor would you expect 'Richard Smith' to come straight from shoeing horses. If you have lived in a rural village you may still recall people saying 'There comes Miller'—an actual miller—or 'Here comes Smith', an actual blacksmith. Their names were still more than an outer label. From the names we bear it is no longer possible to draw conclusions about the nature of the individual. The words that constitute people's names give no clue as to the essential characteristics of the person or persons concerned. The name Smith does not tell us whether the person called by that name is a smith or not, nor can we conclude someone is a miller when we hear that their name is Miller. That is the direction names have taken. The rest of language will follow the same road, and people will have to learn to develop their ideas on principles other than words or phrases. You can draw no conclusions as to the nature of a person from a visiting card with the name Mr Miller. In the same way you will have to get used to the fact that the characteristics of words will not tell you what your ideas about the world need to be.

If you pay serious heed to the urgent needs of the present time you find yourself little understood. If I were to present spiritual science in a way that meets the modern desire for scientific terminology, I would not be doing what I have in fact always made efforts to do: present a subject from all kinds of different angles, sometimes more in their material aspect and at other times more in their spiritual aspect, always remembering the principle which Goethe expressed as follows: 'The truth is never found midway between two contradictory statements.'[55] At the stage we have now reached in our evolution it simply is no longer possible to think that a particular content can be adequately presented by using words to narrowly characterize something. Different aspects must be characterized, and the procedure of characterizing them in words must resemble that of making a photographic record of a tree, for instance, taking pictures of it from various angles. Each photograph will look very different, but one sees something of the tree as a whole by putting them all together.

Read the various courses of lectures and you will see that I have adhered to this principle in presenting accounts from many different angles. If we wish to represent the things human beings need today, things that will serve the progress of humankind, we must accustom ourselves to proceeding in this way. People in certain groups oppose this and wish to continue to use rigid terminology, but human concerns cannot be defined in rigid terms. That is why we now see forms of socialism developing that try to sharpen their terminologies and definitions increasingly, but can only lead to destruction. In Eastern Europe, people think the danger has passed now that the Poles have won; before that the Bolsheviks had the upper hand for a time, but the whole thing has been the most dreadful tragicomedy of human behaviour. The present war between Russia and Poland provides a good demonstration of the extent to which human beings have lost anything that sustains them.

My book *Towards Social Renewal*[31] was genuinely based on the life of modern society, and even its style arose from contemporary social needs. Yet people come and ask you to define your terms precisely, the way words are defined in most schoolbooks nowadays—much to the detriment of education and training. Words have increasingly departed from inner experience, and it has become ever more necessary to draw conclusions about reality from sources other than words. After all, when we hear the name Miller we do not form conclusions about the individual based on that name but on quite different things. Humankind will need to free itself from a fixation on words and judge existence by other criteria.

This had been in preparation for a long time, but it has not always been applied in a way that properly sustains human evolution. This is why all kinds of societies now say: 'We believe in Christ', yet the word they use need not apply to the spirit they say they are worshipping. The point is not that something or other is called Christ and that people have ideas about this Christ. The point is the real nature of the spirit towards whom human feelings are turning. And if one develops a very earthly image of this Christ, if one even undergoes a kind of militaristic initiation during one's training to learn how the soul has to be prepared in order to form an idea of the Christ, if one is shown the image of Jesus the King, seeing oneself and other followers as King Jesus' army, it may happen that having created such a material image of Christ one then gives the name of Christ to quite a different spirit. The truth is that one's soul is then turned towards quite a different spirit who is wrongfully called Christ. This happens a great deal nowadays and it happens in such a way that people sometimes have a curious kind of awareness of it.

Many years ago I had a conversation in Marburg one day with a very well-travelled Protestant clergyman.[56] We talked of how the real idea of the Christ has gradually disappeared from modern theology, of how this modern theology on the one hand uses certain initiation ceremonies to materialize our picture of Christ into a physical Jesus, and how on the other hand certain theologians see Christ only as the 'simple man of Nazareth'. This Protestant theologian, a man who had travelled widely and seen something of the world, then said to me: 'The younger generation of theologians really no longer have the Christ, they really should no longer called themselves Christians or followers of Christ; they really ought to call themselves Jesuits, except that name already has another meaning, because all they are left with is Jesus.' Those were not my views but the views of a well-travelled Protestant theologian.

We have tried to see how the separation of knowledge and belief came about. This separation of knowledge and belief also prevents people from knowing that there is a life before birth, or before conception. I spoke of this yesterday.[57] All that is permitted is belief in post-mortem life, life after death, for this is an idea that requires nothing more of the human soul than egotism. The concept of life before birth, the life we have gone through between our last death and our birth into our present life, can only be understood with perceptive insight; teaching it does not appeal to speculative and egotistical interests. By their very nature, people on earth do not care to know what they have undergone before; for egotistical reasons they are interested in knowing what will happen after death, however. It is easy to preach about the life hereafter, therefore, for this appeals to the egotistical instincts in their souls. You can scarcely preach about life before birth, but must depend on people having a desire to know the truth and to live a life worthy of human beings. This will of course lead us to see education and then also the whole of life on earth in a new light. Life on earth must be seen as the fulfilling of a mission we have been given before we descended from the spiritual world into physical existence.

The impulse that needs to be widely accepted in the outside world, and must also create new social forms, has many enemies. You can guess this from various hidden trends. I want to end today by telling you—as I have to do—of the murky sources and origins of the elements that want to destroy our spiritual science. They are strong and will grow stronger unless souls awaken to a much greater degree than has been the case until now.

You know, and our friends here have fought against it, that the abominable slander has been spread about all over Germany and beyond, of the betrayal of German officers to the Entente by the efforts of proponents of the threefold social order.[58] I have recently been supplied with copies of some of the abominable documents that are widely disseminated at present—fake letters reputedly written by our people, cunningly designed to spread the most dreadful slander, and faked interviews. Their character will be obvious to you as soon as I tell you that one of them concludes with the words: 'D.H. is not in fact part of the Steiner fraternity. He has merely infiltrated the organization to spy on them, to discover their tricks. He has reported his findings to a small group of patriotic people, and the word is that Steiner is committing high treason and is in league with the Entente.' That is just a small sample of the murky work that is being done, and it much more widespread than you would think.

A very nice example comes from someone in this neighbourhood[59] whom I once referred to as 'a swine' in a public lecture—because everything this person is instigating against me simply deserves no better name. They are now spreading calumnies against me in an article headed 'Threefold Order Plagiarized'. This says no less than that a lady had conceived a threefold order some time ago. (The lady was not quite careful enough, however, for she failed to find out from the literature that my threefold order was known before that in certain circles; she gives a time that is somewhat later than the date I began talking to a great many people about the threefold order.) This lady is said to have conceived of a threefold order and to have sent the manuscript to a philanthropic society; it is then said to have gone to Hamburg where the person concerned kept it for four weeks rather than two, during which time I am said to have probably read it and to have stolen the threefold idea. Of course the lady cannot very well say that there is any agreement between the threefold order I present and whatever she had written in her manuscript. She therefore maintains that the threefold idea was plagiarized from her manuscript, but that it has been adulterated. 'He's pinched my watch, though that one looks quite different!' She has now written a book about her threefold order. According to her this consists of 'state, cultural sphere, church', with everything determined by the golden section. So we get a centralized state and within it two parts—'just the same' as suggested by our threefold order! I heartily recommend this work to you; the title is *3:5, 5:8 = 21:34. The Secret of How to Clear the National Debt in a Foreseeable Period* by Elisabeth Mathilde Metzdorff-Teschner,[60] published by the author in 1920. Perhaps we would do well to say that 'We have been working for the threefold order, but really only did this in Mrs Elisabeth Metzdorff-Teschner's name.' That is another thing she expects of us, and she is writing letters to all kinds of people.

All this provides Mr Rohm[59] with his source, and the things he writes are now reaching Switzerland where they are conveyed to the people by every Roman Catholic parish priest. No one of course has the least idea of the actual source. These articles rewrite it all, and people find it quite easy to believe, for the idiocy at the source of it is not apparent. That is the way people work nowadays and they know very well what they are doing. They are deliberately working against the sincere efforts that are being made to serve the true progress of humankind. In Switzerland it is the Roman Catholic parish priests, above all, who republish everything that comes from headquarters run by Mr Knapp[61] and others, everything disgorged from the rubbish bins of Mr Rohm and so forth. I cannot help recalling that until recently there were—and indeed still are—many people, even among anthroposophists, who faithfully subscribe to Mr Rohm's *Leuchtturm*

[Lighthouse] journal. They keep dishing up Mr Rohm's views, keep coming up with one thing or another.

I am sorry, I had to give you some small examples—there are plenty more—so that you can see the methods that are used.

The vigour inherent in anthroposophically oriented spiritual science should give anthroposophy the strength to gain more from words than just definitions, but instead to develop a feeling for the truth. Once you have a feeling for the truth you will find your way, and this will take you in a very different direction from what people generally find comfortable nowadays. It is a road to be sought as I described today. It would be more comfortable in this day and age to talk of other things, rather than refer to the powerful adversaries responsible for the distinction between knowledge and belief, who aim to block the road by which knowledge of the sense-perceptible world can become knowledge of spheres beyond the senses.

LECTURE 9

STUTTGART, 8 NOVEMBER 1920

TODAY we shall start from facts relating to human nature and then pass on to certain guiding principles in world history.

From many different points of view we have already considered the rhythmical alternation between sleeping and waking that human beings experience in every twenty-four hour period. Today I want to examine this alternation between sleeping and waking from a perspective we have been less accustomed to considering.

We know that the human being has a threefold nature. One aspect is the head organization. Here, we have first of all the sensory organism which engages with the outside world. The actual brain organism lies more within. We know of course that this is only an approximate way of looking at these things. We cannot simply divide the human being into different parts and areas. We have to be clear that, while chiefly concentrated in the head, the nerves and senses are in fact present everywhere in the human being. Everything said in this respect applies to the whole human being. We base our characterization on the part where the main concentration lies, i.e. the head. So we have the sensory organism meeting the outer world, and the brain organism situated within.

Now what happens to the sensory organism and the brain organism when we change from the waking state with which you are familiar—perhaps not in depth but at least in outer terms—to the sleeping state? As you know, the sensory organism ceases to be active. The brain organism can continue to be observed in so far as our dream life in a sense shines into our souls. If you consider this dream life you will be able to say that it presents you with a kind of surrounding scenery that in some respects is similar to the outside world you perceive with the senses. It contains images of the external world you perceive with the senses. When awake, we are fully aware how dream life presents us with images that, in a way, derive from the outer, sense-perceptible world. When we then take a closer look at the dream world, considering it in an unbiased way, we find that these dream images are connected—that they relate to each other; they interrelate in a way that is as definite as the interrelations and connections that exist in our waking thoughts, though these tend to be more imageless. It may be said, however, that whereas human beings have full control of the way thoughts are connected in the more imageless thinking of their waking life, and are able to use their will to connect one thought with another, this does not apply in the interplay of dream images. Dream images have their own order and sequence. We are passive where they are concerned. If we then reflect on the way in which dream images succeed each other we find that it is as if the phenomena of ordinary thinking continue in a weaker manner, as if they lack drive and will. Residues of sensory life and also of thought life can still be traced in dream life. It will be evident from everything we discover as we consider our dream life—and spiritual science will be able to establish this beyond all doubt—that the human brain, which in a way is the physical basis of our life of ideas, must have undergone a change from how it was in the waking state. In the waking state our will gives us control of the way thoughts succeed each other. In our dream life we have no such control. What is more, our senses have ceased to act and our dream life only contains images that echo or reflect the life of the senses. The life of the senses has therefore also been weakened.

The question we will ask ourselves today is what kind of changes the human brain has undergone here. If you take an unbiased view you will have to agree that spiritual science is right when it says that the brain acts like a sense organ when we dream. A sense organ receives impressions of the outside world and immediately processes them, at least to some extent. The way a sense organ meets the outside world does not involve an element of will, however. If you consider the way the sense organs meet the outside world, and compare this with the dream state, you will find that when the brain acts as the physical basis of dreaming—take it as a working hypothesis, if you like, that it provides the physical basis for dreaming—it has come to resemble a sense organ. It has become more of a sense organ than it is in the waking state; or we may also say that it is not a sense organ when we are awake, for it shows none of the properties of a sense organ in that state.

Now we do not have far to go to understand what happens in dreamless sleep. Dreams hold a middle position between waking and sleeping. If the brain becomes more like a sense organ even when we are dreaming, it must do so to an even greater extent when we are fully asleep. The way we are constituted as human beings today means that we are not in a position to make use of this sense organ in normal life. There was, however, a time in the history of humankind when human beings were able to use the brain as a sense organ to a very considerable degree. In a way, however, the brain always becomes a sense organ during sleep. We know that, between going to sleep and waking up again, the real human being—the human soul and spirit—is in a wider, external world. We will not take time at this point to consider the nature of this outer world; we merely need to understand clearly that our essential soul and spirit is then in an outer world of soul and spirit. We are unaware of the spiritual and soul constituents of the physical world we see around us between waking up and going to sleep. But in the state which pertains during sleep, as spirit-soul beings we are in the outer world of spirit and soul. Today our human constitution is such that we experience ourselves unconsciously in this spirit-soul environment.

This soul and spirit environment in which we find ourselves during sleep was the essential world in far distant times when the wisdom of humankind originated. An echo of those times is still to be found in the Vedic writings, in Vedanta philosophy—in short in the wisdom that was revealed in the ancient Orient. Looking back to those times we find what those early people of the ancient Orient experienced in their outer environment during sleep. For them, the brain was still very much a sense organ when they were asleep. It was a sense organ, however, which did not permit them to think at the same time as they made sensory perceptions. When the people of the ancient Orient were in the world of soul and spirit, they were actually able to perceive what they experienced during sleep. In a way this was reflected in their brains, which had become sense organs. They were, however, unable to think whilst they were in that condition. They had to wait until they were awake, as it were, before they were able to think the things that they had perceived. We actually have outer evidence that things were how I have just described. You only need study anything that still remains of ancient Oriental culture and you will find that the wisdom of that culture took the form of representing the sense-perceptible universe from a spiritual point of view. Astrology, now a mere caricature, was living wisdom in those times. Most of that ancient wisdom was based on the revelations of the stars, the revelations of the night sky, that is, on things hidden from view during waking life. Human beings experienced these things during sleep. They found themselves in the external cosmos, their souls and spirits experienced their relationship with the heavenly bodies. When they woke up, their brains changed from being sense organs to a state partly similar to that of our own brains—except that their brains were constituted in such a way that when they were awake they were able to remember what they had experienced during sleep. The things they remembered lit up in their minds as instinctive imaginations. As people lived their daily lives in the ancient Orient they were able to deflect their inner attention from the sense-perceptible world around them and focus it on the great illuminating pictures their souls perceived as a memory of their night-time experiences. Those were the original Oriental imaginations. Echoes of them are to be found in the Veda and in Vedanta philosophy and literature.

What image did the people of those times have of themselves? It certainly was not the kind of description of the human being presented in anatomy or physiology today, which is based on sense-derived evidence of our outer form. At that time, amongst all the other things they experienced in this outer world during sleep, human beings experienced themselves as soul and spirit beings. They experienced a cosmos that was soul and spirit, and themselves as soul and spirit within that cosmos. Exactly how did they experience themselves? They perceived themselves as their own exemplar. Please pay particular attention to these words. When an individual living in those times had an illuminating Imagination of what he had experienced in sleep, he saw himself as his own exemplar, and was able to say to himself: 'My ideal exemplar looks like this. This exemplar contains specific and detailed exemplars, as it were, of the inside of my head, of my lungs, liver and so on.' People did not have the experience of themselves that we are given in modern anatomy and physiology, i.e. in terms of organs perceptible to the outer senses. They had experience of the ideal model, the exemplar, the idea out of which the sense-perceptible organs are created. Human beings had the experience of being heavenly and divine spirits—the heavenly and divine ideal of an earthly human being. They were therefore less interested in the earthly human being than they were in the heavenly and divine ideal.

This whole complex of experiences also led to something else. It helped people to realize that they had, in fact, been those heavenly and divine ideals before they were conceived or born as physical human beings. In ancient Oriental times human beings were so constituted that they had the experience of being divine and heavenly human beings, and at the same time experienced themselves as they had been before they became earthly. That is the essential point of ancient Oriental cultures. Human beings experienced what they had been before they entered into physical existence on earth. Their conviction of this was only instinctive, but it did give them the firm assurance that they had existed before they came to earth and had descended from a spiritual world into the world of the physical senses. It is a forgotten characteristic of the ancient Oriental religions that they were very much concerned with life before birth, and presented life on earth as a continuation of life in heaven.

I have already said on another occasion, and from another point of view, that on the whole our era no longer possesses the kind of awareness that belonged to those times. We have a word we use to express the continuing of life after death, the word 'immortality', deathlessness. We do not have a word to express that the beginning of an earth life is not the beginning of life altogether. There is no word similar to 'immortality' that refers to the time before birth.

We ought to have the term 'unbornhood'. If we had that word, and if it were as alive to us as the word 'immortality', we would be able to enter into the state of soul that people had in the ancient Orient.

If you were to put yourself in the state of soul of someone living in the ancient Orient you would be able to see that, for him, life on earth did not merit much attention, for it was merely an image of life in the realm of the spirit. Nor did the people of the ancient Orient take themselves very seriously as physical human beings. The human being walking around on this earth was merely the image of a heavenly human being, and it was this which largely occupied people's minds. The eternal aspect of the human being was a fact that was immediately apparent to those Orientals, for it came to them as an illumination, as I have said. In daytime life, during their waking hours, they had the memory of their night-time life. To picture such a state of soul we have to go back to the ancient Orient.

The great culture of the ancient Orient goes back to far distant times. Any of it still to be found in books, even in the glorious Veda, in Vedanta philosophy, is merely a faint echo. To see the contents of that ancient Oriental wisdom in their pure original form we would have to go a long way back to a much earlier period than that of the Veda. This can only be done with the aid of spiritual science. In that ancient Oriental culture the whole of life on earth was illumined by insight into the spiritual world—an insight that, whilst it may have been instinctive, was also sublime. This culture then fell into decadence. If you take a good look at Oriental culture as it essentially is today you will find that the underlying impulse is still to focus attention on the divine human being. Echoes of this underlying trend are to be found even in Rabindranath Tagore's superficialities. Tagore is entirely immersed in a later, decadent culture but, as I said, the underlying trend is still there in his writings, which in part are of tremendous interest and significance though basically completely superficial. An example are the essays collected in his book on nationalism.[62] When we look to the Orient, therefore, we see an ancient, sublime, instinctive culture with a marked emphasis on life before birth. And we also see the gradual decline of what originally was a sublime culture. The decline reveals an inability to take up the mission of modern humanity, to enter properly into the existence we have between birth and death. In ancient times the people of the Orient received an ideal image of the human being. They saw life in the physical, sense-perceptible world as a reflection of that ideal. This heavenly and divine ideal had been full of life and luminosity. Gradually it darkened and became obscured and all that was left was a shadow image. By now it has faded completely. A shadow image remained of something that once presented itself to the soul as alight and alive, the ideal image the human being had of himself as soul and spirit, part of a whole cosmos of soul and spirit.

However, a certain impotence was also retained in the nature of Oriental peoples. This is something of which we must take special note if we want to live in accord with our age. Orientals were left with a certain inability to observe the human being in terms of image and reflection in the life between birth and death. Orientals had no real interest in this in the past, not even when what they encountered was not a substitute for—an after-image of—divine nature, but something quite different—a human being who was both heavenly and physical. Even today they are not really interested in human beings in their existence between birth and death. It was left to another culture to consider the true nature of the human being here in the world of the senses between birth and death. It was left to a culture which I should like to call the culture of the Centre. Historically this culture of the Centre first appeared during the latter part of the Ancient Greek period. Original Greek antiquity still echoed ancient Oriental wisdom. Later the element began to appear which I will call the culture of the 'Centre'.

The culture of the Centre emerged more from the South and spread through the late-Greek and then, particularly, the Roman world. Spiritual vision was the characteristic of the Oriental culture I have described. The element that arose from the South, spreading through the late-Greek world and assuming its true form in the Roman world—finally becoming the culture of the Centre—came to be a culture based on law, dialectics and intellectual thinking. It came to be a culture not of visionaries but of thinkers. This intellectual culture is especially able to contemplate the human being between birth and death. It went through preliminary stages in the late-Greek period, grew resilient and indeed brutal in the Roman Empire, and was kept alive in the language of Ancient Rome, the Latin language, the language used by scientists right into the Middle Ages. This dialectical and intellectual culture reached its high point at the turn of the eighteenth to the nineteenth century. That was the time of Schiller, Goethe, Herder and also the philosophers Fichte, Schelling and Hegel. Consider the characteristic nature of those great minds and you will see that I am right in what I am saying.

Take Fichte, Schelling, even Goethe. What made them great? Their greatness and significance has to do with perception of the human being between birth and death. They demanded a perception and understanding of the human being as a whole. Take Hegelian philosophy, for example. You will find that great emphasis is put on the spiritual nature of the human being. The spirit is, however, only considered in so far as the human being lives between birth and death: Hegel never considered the pre-birth existence of a heavenly and divine human being, nor any life after death. He presented a historical approach to everything that happened among human beings here on earth, always in so far as they were human beings living between birth and death. You will find nothing about the influx of powers from the world in which human beings live between death and rebirth. It is as if all this had been erased from that great culture, for its mission was to emphasize very clearly that *here*, in the life between birth and death, human beings have soul and spirit as well as a physical body. That culture had its limits, however, in that it was not possible to look up to a life in the spirit. The soul principle that goes beyond birth and death, the eternal element, was given tremendous emphasis particularly by Hegel, but also by all other great thinkers, especially in Germany. Yet they only took account of it in so far as it came to revelation between birth and death; they completely lacked the ability to see into life eternal as it comes to revelation before birth and again after death. When people spoke of a human being independent of the body, they were using an original tradition that had not welled up from their own perception. It was mere tradition. In the intellectual life of Central Europe at that time, tremendous perceptive powers had been developed that focused on the soul and spirit of human beings, but at the same time also on their physical bodies. These tremendous powers did not, however, extend beyond the life between birth and death.

In the West all kinds of new beginnings were emerging for a different kind of life that will evolve in times to come, when a spiritual principle that is free of the body will come into its own in a different way. Let us recall—how did the people of the ancient Orient let the spiritual element enter into their lives? They remembered in the daytime the things they had experienced at night, when they had been outside their bodies during sleep. This will be different in times to come. Today we have merely the early signs, the preliminary stages of this.

During waking life we do not merely experience the things of which we are conscious. Little of what we actually experience at present comes to conscious awareness. The truth is that down below in our human nature we experience immeasurably more than we are able to hold in our awareness. Some people already have an idea of this, particularly in the West. Thus William James[63] spoke of a 'subconscious' or an 'unconscious' because he had an inkling of this, but none of these people have so far been able to achieve full insight. Everything said on the subject is like the babbling of infants, but the idea is there. In the same way that in the ancient Orient the cosmic soul and spirit was experienced in a body-free state, so the time will come when unconscious contents experienced in the depth of human nature will rise up into the awareness of people in the Western world. Imaginations will also arise. Associative psychology as it is practised today is a nonsense, but anyone who has studied the different psychologies of the Western world today can see that it is a preparatory stage.

In times to come something that came to the people of the Centre only as a revelation of human experience between birth and death, will reveal its eternal aspect through the special faculties developed in the West.

Down below exists, especially, what will live in the spiritual world after death. Remember what I have told you about these things on different occasions and from different points of view. I have said that the human head is the outcome of the previous life on earth. The other parts of the human being will become the head in our next life on earth. Those other parts of the human being may be flesh and blood, muscle, skin and bone as we see them today, but in essence they contain the germ of what will be the head during the next incarnation. They therefore relate to the period after death. This connection with the period after death will emerge and come to conscious awareness in the humanity of the future. The early, primitive stages of such a humanity are already present in the West. In future, the inner soul and spirit will be imaginatively perceived, just as the soul and spirit in the world outside human beings were perceived at an instinctively imaginative level in prehistoric times. The difference will be that the revelation of these inner aspects will come to full awareness, whereas the people of the ancient Orient received revelations that were more instinctive and came only dimly to awareness.

What are the early signs of this that we can see today? The first signs are that in these Western regions people are very much inclined towards materialism. In time to come, the spirit will manifest out of physical human substance. Because of this the Western world is tending to become extremely materialistic. That is the cause of the materialism that is predominantly a Western product and, coming from the West, has overrun the Centre and is spreading to the East.

The culture of the Centre is not intrinsically materialistic. We might call it physical-spiritual, because in its view of the human being between birth and death a balance is maintained between consideration of the physical aspect and of the spiritual aspect. German philosophers, Goethe and Schiller, have always given equal validity to body and spirit, as it were. In the West the spirit is a concern for the future; at present attention focuses on the body. Yet everything is in a state of flux in human evolution and this understanding of the body, this materialism, will one day become spiritual understanding—though this spiritual outlook will spring from a quite different source than the spiritual outlook of the ancient Orient, and above all it will be conscious.

So you see—and I have discussed this previously from other angles—the distinctive way in which three different views of the human being are distributed across the globe. In the East, human beings once saw their own heavenly and spiritual image in themselves. In the Centre, human beings see themselves as inhabitants of the earth endowed with soul and spirit as well as a physical body. In the West today, human beings see themselves as merely physical; it is to be their mission, however, to develop faculties out of this physical human body that will be the spiritual content of human awareness in time to come. The early signs of this are already apparent.

The human beings of the Centre are held as in a vice between East and West. The East originally had a very advanced culture but it has fallen into decadence. In the West a great culture is to come, and the first signs are there, but at present people are still entirely caught up in the material world. In the Centre a culture has evolved that, I think I can say, holds the balance between the two. On the one hand we have the clear dialectical thinking of Schiller's letters on aesthetic education, for instance. This way of thinking reaches a point where it does not yet succumb to the superficiality of modern science but still retains a personal human element. On the other hand we have pictures of human society such as those in Goethe's fairy tale of the Green Snake and the Beautiful Lily.[64] This approach does achieve pictures or images, but it pushes them on into spiritual visions.

The people of the Centre have therefore also been given the mission to extend, through direct perception, the insights their particular faculties have given them into the nature of the human being between birth and death, and apply these to the human being as both spirit-soul being and physical-corporeal being, in this way ascending again directly to mystery wisdom. Thus, by developing those same faculties that have rescued soul and spirit, through clear thinking that develops into Imagination, Inspiration and Intuition, human beings rise again to the spiritual world in which they live between death and rebirth. Here in the physical world we will only come to experience the full illumination those faculties can give, once they have been developed, if we consider the problem of freedom. In my *Philosophy of Freedom* I therefore focused entirely on that particular problem.

There it was necessary to use the same faculty, albeit merely to deal with earthly problems, which, when developed further, raises our gaze to encompass the world that lies beyond birth and death.

You see, in a sense the world reveals three stages of evolution: in the ancient Orient an instinctive wisdom, in the Centre a certain dialectical and intellectual life, and in the West today still materialism with the spiritual perception of the future to be born out of it. In the ancient Orient everything depended on that instinctive wisdom. Political life as we know it did not yet exist. The people who presided over the mysteries also set the tone for political and economic life. Greatness for the people of the ancient Orient lay in life of the spirit that developed instinctively. Political and economic life depended on this life in the spirit. The lifestyle of the European Centre did, of course, originally come from the South; its first beginnings go back to Egypt. The way of life that evolved in the Centre reached the point where the state, the political element, was conceived dialectically. Political life—the state—really developed in this culture of the Centre. The life of the spirit became mere tradition. In the West, finally, in Puritanism, for instance, the spiritual element became something entirely abstract, something that could become sectarian, and people let this inform their ordinary, everyday physical lives.

The European Centre therefore provided the soil where, above all, political ideas such as those of Wilhelm von Humboldt[65] were further elaborated. They even assume the marvellous form of 'social community' as in Schiller's letters on aesthetic education, and are presented in grandiose images in Goethe's *Tale of the Green Snake and the Beautiful Lily*, which, fundamentally, represents the idea of the state.

In the West, ideas that have so far developed only in relation to material things, to economics, will one day have to evolve into the threefold social order. The idea of the state has merely been inherited from the culture of the Centre. Woodrow Wilson, once very famous, has written a large volume on the subject of the state.[66] This contains nothing that has originated in the West; all it does, right down to the details of its ideas, is repeat the theories relating to the body politic that have been developed in the Centre. The book has even been translated into German, because in Germany, too, Woodrow Wilson was for a time considered a great man.

It may therefore be said that the idea of a threefold social organism which we envisage has evolved in three historical stages. In the ancient Orient, the life of the spirit gave rise to instinctive ideals. The culture of the Centre was partly instinctive—the idea of the state developed by Humboldt, Schiller, Herder and others who were to follow is half instinctive and half intellectual—with an emphasis on the sphere of rights and on political life. Economic life, as such, really is in the first instance the business of the West. It is the business of the West to such an extent that even the philosophers of the West are really misplaced economists. Spencer would have done a great deal better to have established factories, rather than philosophies. The specific configuration of the West really suits the structure of a factory. There you will find all the things that Spencer was considering.

There is also another way of putting it: in the ancient Orient human beings ascended to the divine within the human being. For them, the human being was in a sense the offspring of the deity, the issue of the divine principle. One can say that the divine, as the Orientals saw it, reached downward, continued on downward as a kind of echo or reflection: the human being on earth was a continuation of the divine exemplar. They saw the divine and spiritual human being above, and the physical human being—as the image of that divine being—in the world below. They merely saw something of the heavenly human being hanging down, as it were, reaching down into the physical world. Later the heavenly human being was forgotten, only a faint idea of it remained in a culture grown decadent, and people no longer had any feeling for something of the divine human being reaching down into the human being on earth.

The people of the Centre are organized in such a way that the aspect of the heavenly human being reaching down from the heights of the spirit has condensed into a kind of closed semicircle, with the physical human being appended to this beneath. A being of divine spirit and physical, bodily nature, a being the mind could entirely encompass, was the result. This is beautifully shown in Hegel's philosophy and Goethe had it beautifully present in his mind.

In the culture of the West, attention focuses on the animal world, on animal nature. Darwin presented a magnificent view of its evolution. At the top is a kind of summit, the human being, though not fully understood, seen as the highest product of evolution. In reality the West considers only animal nature, just as the East only considered the heavenly aspect, the god finding continuation in man. In the West attention focuses on the animal world. This comes to a summit in a creature seen as a continuation of the evolutionary sequence of animals, a kind of super-animal extending beyond animal nature. That is as far as the West has got. The point which has been reached is reflected in Western philosophy. It will develop further, and the people of the Occident will one day give form and substance to the spiritual element from below, just as the people of the Orient received it from above. But in the West it will be done in full conscious awareness. The Centre represents the transition between the two.

When considering realities it feels wrong to speak of an age of transition. Every age is one of transition of course, because there will always be something that went before and something that follows. Yet in the same way that a plant has its calyx in a definite place, with the flowers above and the leaves below, one does get clear divisions and stages in evolution. We can certainly call the time when the great slaughter of the war was in progress, from 1914 onwards, a time of transition, a time that stands out in the historical evolution of humankind. It also was a time when the destiny of the people of the Centre developed in a way that is full of inner tragedy in certain respects. The people of the Centre were faced with a great question: 'How do we find the way from physical life between birth and death on this earth to life between death and rebirth?' Hegel's philosophy soon turned into materialism. The first half of the nineteenth century was unable to answer the question, 'How do we extend the insight we have gained into the spiritual element present here on earth to the spheres beyond this earth?' That indeed is the great question specifically facing us, the question facing the culture of the

Centre. Goetheanism must be developed further. It must develop in the direction of soul and spirit. It must grow out of merely physical human concerns and become cosmic. Anthroposophically oriented spiritual science is attempting to do this. It is a continuation of Goetheanism, extending into the spiritual realm. Goetheanism must be extended and elaborated further to become mystery wisdom.

That is the significant aspect of the signature of the present time. We must understand it before we can consciously take our place in the life of the present, in the work that has to be done today. The Central European element has been severely put to the test. If it does not falter, its task will be to deepen its perception of human existence in the physical, sense-perceptible world; a perception in which the spirit is still present in the physical, sense-perceptible world. That will have to be the foundation for developing a mystery wisdom, with the same clear intellect as used for understanding the physical, sense-perceptible world. The European Centre therefore must, or ought to, come to understand very clearly how a balance is achieved between the three spheres of culture, politics and the economy. Other regions will then simply follow suit. Here in the Centre people would be utterly remiss, however, if they refused to wake up, if they ignored the great necessity that has arisen—to grasp and put into action the impulse for a threefold order of the social organism.

The European Centre is held as in a vice between East and West. Today it lies prostrate. Out of the very darkness of despair it has to find its way to the light.

In the next lecture we will talk about what is to happen before the middle of this century. I shall speak to you about the Christ appearing before the middle of the twentieth century, as I hinted at in my first Mystery Play. For the moment let me just say that this reappearance of the Christ is closely bound up with our understanding of the threefold nature of the whole of the cosmos. It will come about in so far as the Centre necessarily turns its attention on the one hand to the instinctive spiritual culture of the East, a culture grown old; and on the other hand to the West. It must turn its attention to the West with a thorough understanding of what is in preparation there in a culture that is still materialistic today, but whose materialism holds the seed of a spirituality of the future. The culture of the Centre must take its place in the middle; it must find the energy and the strength to take its place there and point the way.

It causes me great and heartfelt pain that souls are not open today to receive words relating to the necessities I have spoken of. It causes me pain that people wish to remain asleep, wish to relinquish themselves; that they shrink from the great tasks that now face us. We must look to the East and look to the West and understand what is in progress there.

It has to be clearly understood that Western culture is in its initial stages. We can see that this is most immediately apparent at the point where economic processes sprout from technological processes, if I may put it like this. A very typical example is the ideal once conceived by an American, an ideal that is bound to come to realization in the West one day. It is a purely ahrimanic ideal, but a high ideal nevertheless. It consists of using the vibrations generated in the human organism, studying them in great detail and transferring them to machines so that if someone stood by a machine it would intensify even their least vibrations. The vibrations of human nerves would be transferred to the machine. Think of the Keely engine.[67] It did not succeed at the first attempt because it had been developed largely by instinct, but it is something that will certainly be realized one day. Here something arises from the crude mechanistic, material world that points to what is to come—material mechanics linking up with immaterial, spiritual elements.

In the East, on the other hand, the old spirituality is increasingly falling into decadence, into decay. It is rotting away. Our experience of the East is such that we may certainly say that the human being once perceived as a heavenly, spiritual being has come to seem old and senile. This human being still has no understanding for the things of the earth, for the things that also clothe human beings on this earth. The West understands earthly things only, the East has no understanding of them. Because of this, the heavenly element has grown completely senile. It is always a great mistake not to pay proper attention to the way in which the spiritual element still has to be won from mechanical genius, the mechanistic materialism of the

West, how the spirit will need to be intuitively derived from a science that is, as yet, very much subject to Western materialism. In the same way it is a great mistake to cast sidelong glances at the East and to try and bring the spiritual life of the East to the West, in this day and age. The Theosophical Society based at Adyar used to do this and perhaps still does in its antiquated ways. Looking to the East, nothing one finds there has anything in it that relates to modern life; it is something grown old, and has to be studied as something historical that has grown old—something of no significance for the present.

In the West, if I may put it like this, we have Keely and his engine as a rough, crude mechanistic forerunner of a future culture. The last gasp of the East's spiritual senility on the other hand may be seen in the work of Tolstoy. There we see a concentrated form of something that has once been great and is now completely decadent. This is an interesting phenomenon but it does not have the least significance for the present. Much has been wiped out with the events that happened from 1914 onwards, including that last flame of Eastern senility flickering in Tolstoy. Before the war it was still possible to speak of Tolstoy as a writer of significance for the present. The war has put an end to this and Tolstoy is no longer of significance. It is definitely out of date to speak of Tolstoy as though he were of significance today. And we must take care not to cast any kind of sidelong glance in the direction of the East, of the ancient East, and at the things that have in a way grown senile and concentrated themselves again in an individual such as Tolstoy. We must take our stand for the mission of the present. We can only do so if we grasp through our own intrinsic nature the impulse for a threefold order of the social organism. The decaying East has created a symbol, as it were, in world history—or we might say a symptom—in having in Tolstoy a kind of last gasp, full of inward endeavour, and yet impotent. The West on the other hand has produced Keely with his engine as a first forerunner. Tolstoy showed how the old Oriental culture had grown completely luciferic; Western culture is still entirely under the sway of the ahrimanic element.

This is what we must grasp today. On the one hand we must be wary of past elements reaching across from the East, of past elements from the East in someone living in this century; and on the other hand we must be wary of what is only in its infancy in the West. If we fail to grasp this and fail to perceive the true nature of these things we do not belong to the present age. Someone belonging to the present age may of course be English, French, American or Russian—humanity extends beyond all geographical boundaries today. It is important, however, to consider the old geographical limits because of their role in the historical evolution of humankind. Behind us lies a history of humankind that occurred in three stages—Orient, Centre, West. Before us—and this is something anthroposophically-oriented spiritual science must really stress—lies the time when we will be solely *human* beings, holding in balance within us simultaneously the East, the Centre and the West. Anyone truly alive today—and this includes anyone who is Asian—is capable of bearing all three within them. The people of the Centre need not limit themselves to holding the Centre within them. They must gain inner experience of the historical East in its decadence and the historical West which is in the ascendant. And Americans can hold East, Centre and West within themselves if they attend to mystery wisdom—they actually need it more than most—and raise their merely economic thinking to one that is political, that is cultural and spiritual.

That is what we must say today when we try to define the tasks which human individuals, attending to the great needs of our era, should recognize as their inmost concerns.

Lecture 10

STUTTGART, 14 NOVEMBER 1920

So far we have considered from various angles the forces in human evolution which we must come to know if we wish to arrive at a proper understanding of current events, and above all the root causes that have led to the disastrous situation we face today. Only then can we find the right way of taking effective action ourselves, and work for the real progress of humankind. Unfortunately, far too little attention is paid to the changes which those forces active in human evolution have undergone in very recent times, compared to times that were not so long ago, relatively speaking.

Perhaps I may again be permitted to take the great catastrophe that has happened in recent years as my starting point. This will lead us to the event I briefly mentioned at the end of my last talk; the distinctive Christ event, to which I have referred a number of times, which belongs to the first half of the twentieth century.

Taking a genuinely unbiased view of the disastrous events and of their consequences that continue into the present and will go on making themselves felt for a long time, we cannot fail to notice how much the whole fabric of the destiny of civilized humanity differs from earlier times. Let me point out immediately that large numbers of people, including those in authority, have not yet become fully aware of what has come upon us. They have been, and still are, acting in accord with the demands of earlier times and their actions are not at all in accord with the needs of the present age.

Again let me give an example, just by way of introduction as it were. We have a—well, they call it this—a 'war' behind us; a much greater war than any known in historical times. We have seen that at the time when the war started, something like a spectre from prehistoric times lived in people's minds, and that this spectre is still present in the minds of people today. We have seen that this spectral idea continuing into the present from prehistoric times has given rise to opinions about what should have happened, leading on to all kinds of actions and measures; people had no idea that something fundamentally quite different was going on from their conception of these events.

Like past wars, the last one was a conflict between battling human beings. Unlike past wars, however, it involved energies and forces of quite different origins from those at work in earlier wars, which derived from things intrinsic to human nature. We have seen a tremendous development of technology in recent times and the coming of this powerful technology has changed the whole fabric of human destiny. This radical change determined the course of events in recent years. Yet there has been no corresponding change in people's ideas.

Let me present the major aspects of this. During the period preceding this catastrophic war, the human technology which had developed in modern times had reached a significant stage. Although people had no real awareness of this, human work—labour—had taken on quite a different form to what had gone before. You can get an idea of the different forms it took if you consider one of the fundamental aspects of modern technology, let us say coal-mining, in different countries in the civilized world. The amount of coal produced in the mines relates to the amount of energy gained by processing it; energies merely channelled and controlled by human beings that will then work more or less independently. I would say, therefore, that in recent times human work has come to consist more in stepping back and directing the work actually done by machines.

Surveying the situation one will find, for instance, that during the period preceding the outbreak of war, 79 million 'horse-power years' of that kind of energy were produced in Germany. These were energies controlled by humans, but in fact derived from coal-mining. They did not arise from something that human beings allowed to spring from within them, but from completely external actions and processes. The energies expended in work are measured in units based on the work a horse does in a year. During the period immediately preceding the outbreak of war, Germany was producing 79 million horse-power years of coal-derived energy per annum.

What does this really mean? A very superficial comparison with the population figure for Germany shows that on average every single individual in Germany was metaphorically accompanied by one horse. This means that the work accomplished in the technological field by the inhabitants of Germany was equivalent to every individual having one horse work for them all year long. The population figure was approximately 79 million, and the energy used was 79 million horse-power years. The work done by machines, all kinds of machines, therefore had the same effect as if every individual had a horse to work for them. When war broke out, the potential was there for that much work to be done. A large proportion of this work was then used for the purposes of war, with the result that, as it were, the purely technological effect of 79 million horse-power years was brought into action at the Front.

Now let us take some other figures. For a start I will just add the fact that in 1870, a year when people—rightly—consider that another great event[68] took place, instead of 79 million horse-power years or energy produced, the figure was six million and seven-tenths, a very low figure compared to the output of human energy. Six and a half million in 1870, 79 million in 1912. Clearly this means a complete change in the human situation.

Let me give you some more figures. During the time preceding the disastrous war, France, Russia and Belgium together had 35 million horse-power years available. Great Britain on the other hand had 98 million. Due to the geographical position of Great Britain, those 98 million horse-power years could not immediately be fully brought to bear in the war zone; it took some years until this was achieved. When war broke out, therefore, not only were human beings opposing each other, but 79 million horse-power years had been introduced into the front lines by the Germans, or a total of more than 90 million by the Central Powers [Germany and Austria]. A large part of those energies were of course used in the war industries and therefore reached the front lines indirectly.

Those energies were opposed by Great Britain's 98 million horse-power years, which could only be brought to bear gradually, and a total of 35 million available in Belgium, Russia and France. You can see now that it would be quite correct to say that, basically, the contribution made by human beings only gave a preliminary result in the initial stages. The army's General Staff could order the troops to march; that could be planned and projected in a certain way. But when the front lines had been established for some years, the actual confrontation was between the horse-power years that technology had produced—and these were quite independent of human beings. Thus the relative size of something that had really been taken out of the sphere of human activity determined the fate of this phase of development. If you now take the following and add it to what has been already said you will see that forces independent of humanity, particularly the most recent achievements of technology, have been responsible for the events that occurred.

The efforts of human beings, limited of course to a channelling or directing function, or at best to preventing certain things from happening, unleashed objective forces that entered the field of battle: some of these forces could overcome others based on objective laws that had nothing to do with human beings. Add to this the fact that the United States intervened in the process. At the time when the other countries were able to mobilize the number of horse-power years I have mentioned, the Americans were in a position to mobilize 179 million horse-power years. This gives you the relative figures for energies mobilized through technology; energies completely independent of anything human beings are capable of producing. Indirectly they are of course connected with ideas conceived by human individuals and so on. The things people thought up were, however, channelled in this direction, and in the war objective force met objective force, this deciding the issue in the final instance. In the most recent times, very different from the course things took in the past, human beings and human actions have surrendered human destiny to the powers at work in the objects they themselves had produced, becoming dependent on earthly productivity, on all kinds of factors that do not lie within their own skins.

This reveals one of the characteristic features of the present age. I have merely given the most striking example to illustrate it. Yet the things that happened there on a gigantic scale—we cannot merely say on a large scale, because it was gigantic—happen on a smaller scale in everyday life and our overall destiny, for we have been delivered up to the products of technology. In 1912 a point had been reached in Germany where fertile human brains had created something outside themselves that did the same amount of work for every individual as a horse would have done. That is the characteristic feature of modern civilization, and we must take a good look at it.

What is alive in those forces that human beings have objectively created outside themselves in modern civilization, forces that work for them every day and determine their destinies? Looking at these forces and the way they influence human destiny we perceive the power that we have come to call the power of Ahriman. Ahrimanic powers are alive in these things. Looking at it like this, you will have to admit that the power of these ahrimanic forces has increased at a tremendous rate. Just consider those two figures I gave: in 1870, six and a half million horse-power years were available in Germany—not very much per human being—compared to 79 million horse-power years in 1912. That is the sum total of these things influencing not only economic life but the rest of our life as well. So you see what goes on in a world constructed by human beings that is quite independent of what really lies in human nature. These forces are completely at odds with everything that came into effect when people faced each other as human beings, in the battles fought in the ancient Orient, for instance. Only luciferic forces were involved in those, and this still held true when the Tartar hordes invaded Europe, for example. We often do not realize how different the world has become for modern humanity and how quickly this has come about, relatively speaking.

Anthroposophically oriented spiritual science is called upon to consider the full implications of such a fact. So far I have merely described the outer aspect. We begin to see the inner aspect when we consider the powers which are active in this. In the past they were luciferic powers and now they are ahrimanic. Human beings find themselves in the middle between the two. First of all, however, we must gain a definite idea of what we mean by 'ahrimanic' and 'luciferic'.

Consider what went on in human soul life in past times when the great struggles in which human beings were involved were largely determined by luciferic elements. At that time people looked at the phenomena in the world and, as you know, perceived in them a certain number of elemental beings—let us say demonic spirits. Materialistic scientists call this the 'age of vitalism', saying that people introduced all kinds of water sprites, gnomes and so on into the phenomenal world. We know that spirits are indeed active in the phenomena of nature. Today, people see only dry-as-dust, prosaic natural phenomena. In the past, people perceived the spiritual entities, the essential spirit of natural phenomena. This is called superstition, nowadays. That is the view taken by the present age. We know, however, that the people of the past used those names to describe something real, which they perceived when they looked at natural phenomena. They saw elemental spirit in everything nature revealed to them. Thus we may say that however instinctive, dim and dreamlike their conscious awareness may have been, some illumination was received as to the nature of those elemental spirits.

During succeeding eras, perception of the spiritual essence of natural phenomena no longer came clearly to mind when people looked out upon a natural environment that did not depend on them for its existence. The modern intellectual approach that we call the scientific outlook came into being. This only concerns itself with forces that can be abstracted from nature and made comprehensible through abstract ideas, in short, that can become the content of the human intellect.

I would say, however, that without people being aware of it, a completely new world developed in a relatively short time—just take the period from 1870, when 6.7 million horse-power years were expended in Germany, to 1912, when 79 million were expended. This is a world that did not exist before. These forces are present in the human environment, and in major events like those we have seen in recent years. Human destiny actually depended on them, just as formerly it had depended on natural phenomena. These forces and energies also exist and take effect independently of human beings, just as the forces of nature are independent of human beings. Demons, elemental powers are active in them, but their effects on human beings differ from those of the elemental powers that human beings perceived in natural phenomena in the past. Back then, people would look at the phenomena of nature and say: 'Elemental spirits are at work there.' That had an effect on their conscious awareness; the soul came to an accord with the phenomena of nature, and united its consciousness with them. Today's 'enlightened' minds consider it superstitious to look for spiritual powers in natural phenomena. They have not the least idea that demonic spirits are active in the whole world of technology created by the human race. Nor will they find it easy to see this, because those powers are acting on the will—and I have often told you that the will is asleep. They work at an unconscious level, taking hold of the unconscious human mind. In the past, human beings had at least some awareness of demonic powers. Today demonic powers are restively stirring in all products of technology;[69] their activities extend into the sphere of the human will, but human beings are not yet inclined to accept this. In the first place these things act at an unconscious level and in the second place people feel it would be superstitious to say that demonic spirits are active in the machines they have produced. They are active nevertheless. The spirits perceived in the phenomena of nature in older times were luciferic by nature; the spirits active in machines, in all products of technology, are ahrimanic by nature. Human beings are thus surrounding themselves with an ahrimanic world that is growing completely independent of them.

You will perceive the trend in human evolution here. From a luciferic world that still influences their conscious minds and determines their destinies, human beings are drifting into an ahrimanic world. And at present this is happening at quite a fast pace. This ahrimanic world acts on the human will, and the intellectualism of modern science does not enable people to gain immediate conscious awareness of the will. The great danger is that the ahrimanic world will take hold of the human will and human beings will completely lose their bearings among the demonic powers that are present in the products of technology.

In Eastern Europe contemporary thinking has led to a desire to militarize the economy and make it into a vast machine. Even human beings are trained to be like machines, with human labour made into something quite separate from the human being. Behind it lies the will to call forth will-demons, for it is into their sphere that people are sliding in Eastern Europe.

The road from the luciferic to the ahrimanic sphere—yes, it has to be said that the course of human evolution is heading in that direction. We find ourselves at present, basically, at the very point of departing from the luciferic and slipping into the ahrimanic sphere. Luciferic elements are of course still present in many ways. The ahrimanic element is taking hold of people. The luciferic is more alive in feelings. The ahrimanic takes effect through the human intellect and comes to realization as it takes form in the products of technology.

And now the Christ event, which we may expect during the first half of the twentieth century, enters into the situation to give human beings their bearings. The nature of this Christ event will be that more and more people will have objective experiences revealing to them the etheric Christ walking upon earth, the Christ who embodies the power that once walked on earth in the physical form of Christ Jesus, but now does so at the etheric level. If people come to know this Christ power, if they let it enter them, they will find the right way of allowing the necessary advent of ahrimanic powers to act upon them. The great problem of our time is that people slip into the ahrimanic sphere without being sustained by the power of Christ.

We are speaking of something very concrete and positive when we refer to this event that enters human evolution in the twentieth century. In my first Mystery Play I referred to this as the reappearance of the Christ. And it is possible to perceive what will happen in human souls when they direct their lives to meet this Christ event.

The other day, in a public lecture,[70] I suggested that the scientific thinking of the West, which is greatly lacking in coherent vision, reaches its limits when it comes to perceiving the nature of the human being. Science mostly grasps and categorizes the non-living world, developing theories about both the non-living and the living world. Darwinism does not go beyond the evolution of animals, however. The human being is then placed at the top of the tree, but this theory really does not include the human being. It is unable to perceive the nature of the human being.

The same applies to a grasp of social concepts. I have shown that people really run along fixed lines when working in this field; their approach is ahrimanic, technical, and they do not go beyond this. They note it all down in the manner of accounts, the tallying of credits and liabilities. They fail, however, to consider the people involved. Those people want their dignity as human beings to be recognized, but no bridges are built between management and workers. Practical life fails to take account of the actual nature of the human being.

On the one hand all this is still more or less theoretical today—though perhaps we should not call it theory but rather empty theory, impotent knowledge. On the other side we have something that is much to the fore in the social sphere today. The things not bookishly accounted for are today making themselves felt in strikes and revolutionary movements. They show themselves in life and they arise out of the work done in industry, in commerce and so on, just as much as all kinds of goods result from industrial production. It is merely that this element, which is now causing unrest among the people, has not been allowed for in the accounts. But it makes itself felt in real life.

I believe it is true to say that very few people really think about these things, as I also discussed at that recent public lecture. The nineteenth century clouded people's minds. During the eighteenth century some people, certainly the more radical thinkers, did begin to get an idea of what was coming. The nineteenth century brought events that caused widespread confusion. Pierre Bayle[71] made a very peculiar statement in the eighteenth century. He was an eighteenth-century forerunner of nineteenth-century materialism. His statement went as follows:

> Nations will know honour and dishonour, ambition, egotism and so forth, but there cannot be a nation state in which Christian attitudes play an active role; it is possible to have a state system in which the old heathen virtues and vices play a role, but there can be no such a thing as a Christian state.

Those were the words of Pierre Bayle, a radical materialist, and there was more truth in his words than in those of the nineteenth-century idealists. The latter thinkers pretended to themselves that nation states might be Christian. The truth is that they were not. Consider the Christian beliefs of the Middle Ages, for these were foremost in Pierre Bayle's mind. Those beliefs were based on a denial of this earth. It was considered a virtue to rise to a life that was not of this world. The life that developed during the eighteenth century was on the other hand largely concerned with earthly matters. 'There can be no such thing as a Christian state', Pierre Bayle said, and that in fact was the truth. People were lying when, in the nineteenth and early twentieth century, they pretended to themselves and others that the modern states which had gradually evolved possessed the Christ spirit. They cannot be Christian. Something else emerged from this, however. Whether they stood in the pulpit, or heard the words that were spoken from it, the people felt utterly convinced that they were true Christians. In the same way people going to their work in government offices, putting on their medals or using the titles conferred on them by the state, imagined themselves to be Christian. They were not in fact, for they held those very positions by virtue of not being Christian. People became accustomed to living a lie; they got out of the habit of seeing the truth about life's most important aspects. The result was a nebulous atmosphere where it was not even possible to develop an unbiased view of the progressive ahrimanization of the world.

There has been a lot of talk about the campaign of lies of recent years. Yet in all the most important respects people have actually grown used to such campaigns. What reason is there to tell the truth now about the lies that were told during that disastrous war? After all, during the nineteenth century people relinquished the habit of wishing to know and inwardly experience the truth about life's most important concerns.

It is uncomfortable to face up to these things, and the problem is that people do not. Apart from anything else, therefore, modern people are in the difficult position that has arisen because of their inner untruthfulness. This atmosphere will cause a particular mood to develop. Until now it has been mere theory in many respects, merely something we know, that human beings can no longer get through to each other; that the nature of the human being cannot be understood and that this failure to get through to others is also having an effect in the social sphere. All this will become engrained in the human soul. The influence of technology's outward products on the will is going to rise up from the unconscious and impinge on the conscious sphere. There will be no conscious awareness of this, of course, for it is at a subconscious level, but a mood will be created. This mood will emerge more and more over the next few decades, or rather years, taking hold of large numbers of people. You will be teaching children in your school and you will notice that these children express feelings that their elders never had. Something like this has existed before in various eras, but it will happen to a much greater degree in the near future. Profound spiritual insight into the present age will be able to discern what is unfolding from the depths of the souls of these young people. A great longing is going to come, a kind of longing for something that is lacking. Initially, theories could not be developed that embraced the true nature of the human being, and in the social sphere it proved impossible to include human gifts and talents in society's stocktaking. All this will condense into feelings and emotions. There will be people—we shall find them among the young in the next generations—who will feel like this: 'Well, here I am. My form is different from that of other life forms around me. I do not look like an animal, an ox, a donkey, a weasel or an eagle. I look different, but I do not know what it is that looks different. I do not know what a human being is; I do not know what I am.' Despondency and neurosis will invade the souls of the coming generation. This will be the mood of the era that teachers will perceive as they give their lessons. It will be a mood that spreads far and wide. People are so superficial nowadays that is difficult to talk to them about these things. To show what I mean let me remind you that during the eighteenth century people who had some perception of the soul of that age spoke of a 'Werther fever'. Goethe wrote his *Werther*[72] out of the whole mood of that age. Then another novel called *Siegwart*[73] appeared. This had been written in the 'Siegwart fever' of the second half of the eighteenth century. Those were the moods of the times, but they only affected a limited number of people. A mood we can describe as follows will, however, arise in the souls of vast numbers of people: 'Well, what am I as a human being? What kind of life form am I as I walk around on two legs? I have a science that I have taken to great heights; have a life in the social sphere—yet both of them do not touch on the reality of what I am.' This mood will be the great question mark of the age, a question mark as to one's own nature as a human being. It will prepare the eye of the soul so that it perceives something difficult to describe, but will nevertheless arrive as the new Christ event. Out of that longing, the power will arise to see the Christ made evident. The distress of outer circumstances will become an inner distress, an inner need of the soul, and out of this inner misery is to be born the vision of the Christ who walks supersensibly among human beings, the Christ whom they will need to follow to prevent themselves from sliding disastrously from the luciferic into the ahrimanic sphere.

What use would all our science be to us if it could not help us gain a really tangible and direct grasp of human life? We have to understand that human beings as they are today already have a whole succession of earth lives behind them. We live through repeated lives on earth. In earlier earth lives we ourselves have seen elemental powers at work in the phenomena of nature. We have brought the fruits of those earlier earth lives into our present life. Back then we knew that nature spirits around us determined our destines, and that we bore those spirits within us. Today we consider nature entirely with the intellect, with the head, and the products of technology we ourselves have produced are considered in the same way. All we see are the contents of our intellect. But out of the many earth lives we have lived through—though we refuse to acknowledge them nowadays—something is restlessly astir in us that I have called a great longing, a longing for something lacking. We have been human beings who looked at nature and saw the spirit; this enabled us to develop an inner feeling for the true nature of human beings. Now we have a science and a social awareness that do not encompass the human being. Our past vision of the world around us has given us the potential capacity to feel ourselves human beings. Today we look into a nature empty of humanity, we do not penetrate as far as the human being. This will create a great need in the souls of people in the decades to come. This need in the souls of people is a positive power, and out of this positive power will be born the ability to see the Christ.

The old way of approaching the Christ has been destroyed by modern theology. Our modern theology has made the Christ into the 'simple man of Nazareth'. Surely, it will be impossible for human beings to relate to the Christ event today unless there is a renewal of life in the spirit.

The Catholic Church knew very well why it did not want the masses to have direct access to the Gospels. In theory, Catholics are still not permitted to read the Gospels. The Albigensians, Waldensians and others who would not accept this were, of course, declared heretics because the church was only too well aware what would happen if the Gospels became available to the masses. In the first place there are four Gospels. The divine spirit reveals itself to humankind in those four forms. You cannot, however, intellectually present an event in four different ways as it has been done in the Gospels, for then contradictions inevitably arise. The moment you deny the Gospels their reality, considering them to be the products of the human intellect, they inevitably become contradictory, full of contradictions. What has emerged, therefore, is the complete destruction of any vision of the Mystery of Golgotha.

Again people live with the lie that they are supposed to remain Christians, and yet its very wellspring has been stopped up and brought to nought because modern theology no longer has anything in it that is Christian. To regain Christianity we must acquire a new spiritual vision. We must regain access to the treasure we have garnered in our souls and carried with us through many earth lives.

Our present life, the life we now find ourselves in, is also the starting point for future earth lives. The abstract thinking we use for mathematics, and the various moods of which the soul has tangible experience, are the inheritance of former earth lives. Everything we absorb of the outside world in this earth life will form the germ for faculties we shall have in future earth lives. In the past, people came to see elemental spirits active in the world of their natural environment. When we were on earth before we looked at nature and gained an impression of elemental spirits; this is now part of us. Today life is largely determined by what the 'horse' beside us is producing, by technical things. This enters us. Unless we do something about it, this will create the basis in us for our future lives on earth. New demons, the ahrimanic demons, are active within this. What a way of preparing ourselves for future earth lives—to deliver ourselves up to ahrimanic powers! Machines create something in us that will be our life of knowledge in future earth lives. The roar of cannon fire at the Front, what lived in those machines of war, is something we incorporate into ourselves. And so, in our next earth life, we will seek to be unconscious really. But human beings are more than just intellect; they also have feelings and sensibilities. These faculties will have to cope with everything entering the soul from the products of technology, from machines.

And another feeling arises, one I have not yet described. I was speaking of the longing for something that is lacking. At the unconscious level the soul creates something out of the products of technology, out of the ahrimanic powers, and this rises up, reacts upon higher levels, entering our awareness in the form of thoughts and ideas. What arises takes a form that resembles fear. You will find that children you teach at school in the years and decades ahead will show this longing for something that is lacking, and also an indefinite but nevertheless tangible fear of life. This will show itself as a form of nervousness, in a nervy, fidgety character—you will find this very evident. The first signs of it are already to be seen today.

The only way to counteract this is to fill your soul with the one thing that will give strength, a strength the earth itself cannot provide. This is the strength that has come to the earth from outside it through the being of Christ who is now to appear again. It is a power we cannot gain from the earth. The earth provides technical power, the 79 million horses by our side. We must develop in our souls the strength that comes from the Christ. Otherwise we shall be filled with nothing but the power provided by the products of technology in our next life. The only cure for the nervousness that will inevitably manifest in the coming generation is to prepare ourselves for the Christ event that is to come in the first half of the twentieth century.

Our age should not be described in external fashion, but should be characterized in terms of feelings paramount in human souls. The thing of outstanding importance in our era would be for human beings to develop insight into what lives in human beings. As a rule only external aspects are described. People like Paquet,[74] for example—a writer who has travelled in Eastern Europe—are in no position to give a true picture of what goes on in the hearts and minds of those who are already experiencing a great deal of the future; such writers can do nothing but describe things externally.

If spiritual science is to come alive in us it must be able to provide us with insight into the sphere of feelings and sensibilities. You do not know what life truly is if you use abstract terms to speak of the Christ event that is to come; you only know what life really is if you speak of human souls progressing toward an encounter with this Christ event, partly with longing in their heart and partly with fear.

How can people as they are today grasp such things as the way the fate of that disastrous war was sealed by ahrimanic forces, quite independently of anything human beings were able to do? It was entirely determined by elements conceived in human brains that then became objective forces. It is impossible for modern people to get the right picture and assess the real effects of these things—unless they take account of spiritual science. Just think what it means that 79 million horse-power years of work were produced in Germany, 98 million in Great Britain, 35 million in Belgium, France and Russia, and finally the 179 million produced in the United States were added to this! We are speaking of something that takes no account of human nature, yet these are the factors that actually determine human destiny today. Human beings have wholly surrendered themselves to something that is no longer human. This shows in a new light the thought that human insight stops short of understanding the human being. Human beings remain limited to the non-human realm; even in the social sphere they will be restricted to understanding only non-human aspects unless they find the bridge that leads to the reality of human nature. This is how they fulfil their destiny today: they make their destiny partly dependent on elements that are no longer human; they produce elements which will partly determine human destiny, and yet human beings themselves will no longer be able to influence these elements. We ought no longer to be speaking of the courage, of the brilliance of the General Staff and the like when we want to speak of the way destinies are shaped, but of the ratios of horse-power years produced in different countries. In speaking of human destiny we now have to ignore the human being! People will need tremendous strength to rise and face this human destiny which is determined by non-human elements; and to call for the destiny of the human race to be determined by human beings again! This can only happen of course, if people allow the power of Christ to enter them. This power of Christ is approaching and it will restore them to their human powers. We can only become sure of ourselves as human beings if we walk the road created by the whole of technology, but do not let our lives be governed by technological products; if we become able to behold the Christ power that can permeate us and can overcome all these technological instruments.

These are the lessons to be learned today. These are the words that teach us how to prepare ourselves for the Christ event. All the mediocre stuff that makes up the bulk of modern literature, all the empty talk one hears nowadays, will not help humanity to progress but only regress. Progress can only be achieved by things we draw up from spiritual depths. And we will make no progress at all unless we become fully aware again of the seriousness of such matters. We have to understand that humanity has created a completely new world around itself today; human beings themselves have developed the energies and forces that now shape their destinies. This certainly does not only mean the events of war. Step outside and you will see the factories that shape our destiny in ordinary life. This is not restricted to the consequences of 1914. Ahrimanic forces produce those smoking chimneys, in factories where the human being no longer counts.

Walking on past the factory we come to the Church. The Church still has its traditional message but this has grown abstract. It no longer has relevance in ordinary life. The Church concerns itself with things that have no application in practical life. All this is luciferic, just as the things that go on in factories are ahrimanic.

The dreadful thing connected with the destiny of modern humankind is that in the places where people speak of things of the spirit, all ability has been lost to relate to real life. In a public lecture I recently gave[75] I spoke of American preachers coming to Switzerland and other neutral countries and saying something like this: 'The League of Nations must be created, for it will be a great blessing for humankind; but it cannot develop from the ideas of statesmen. It will be necessary to win people's hearts instead, so that they believe in the League of Nations.' Anyone with an unprejudiced mind will realize that these gentlemen make very fine speeches; but if that suffices for us, and we find it sufficient to praise the beauty of those speeches, we fail to understand the signs of the times. The sweetness of such honeyed words does not actually reach people's hearts, hearts full of anxiety about the economic situation, with no bridge between their reality and the words of old religious confessions. You cannot use these words to create a League of Nations, nor can you do so with the words uttered by Woodrow Wilson,[66] Clemenceau[76] and others. What matters today is that the two must be brought together, life must be imbued with spirit, and life must progress toward connecting with the spirit. The Christ spirit from beyond the earth assumed flesh in Jesus, a human being, uniting itself with the physical world. The Christ who will come during the first half of the twentieth century will not use the language of abstract religious confessions—oh no! He will use the language of everyday life. People who are continually looking only to unearthly mystical heights for moral uplift will not understand this language. The Christ will, however, be speaking of the spirit even as He speaks of everyday life. The spirit will unite with everyday life in the same way as the Christ spirit came from spheres beyond this world, beyond the world of the senses, and united with the physical human being Jesus. Such is the new understanding we must gain for the Christ event; otherwise we shall not be able to appreciate its true value when it approaches humankind.

A question we may ask already now is the following: How will those who preach official Christianity respond to the Christ event when it occurs during the first half of the twentieth century? If we understand the Gospels rightly, they are actually exemplary for us here. The Gospels speak of 'scribes and Pharisees'. We would be wrong to place Adolf Harnack[77] among those who testify to the Christ; we only judge him rightly if we follow the example of the Gospels and place him among the scribes and Pharisees. And there are others of the same kind who must also be counted among them. It is essential to judge these things in the right way. We must get to the truth! The materialist Pierre Bayle said that a state could not be Christian, that states might know honour and dishonour, ambition and egotism, but that a Christian state was an impossibility. A social community in the name of Christ *will,* however, be possible, providing we do not insist on a political state but rather establish an independent life of the spirit that can be Christian through and through. And this independent life of the spirit will be able to illumine the sphere of life where government and nation states exist, a sphere that simply cannot be Christian in itself. The result will be that an economic life based on associations can develop; though this, too, cannot be Christian in itself. The people who are involved in it will be Christian, however. They will be filled with the Christ impulse. What we must do is to let people embrace an independent life of the spirit. Then it will be possible to make the whole of our social interplay Christian.

First of all we need truth, however. We shall not prosper with lies. These are the things we must come to acknowledge today, inscribing them deeply in our hearts. If we fail to do so we will be siding with the people who follow Spengler[41] and believe that we will inevitably become barbarians. Yet it will not help either if we too superficially judge Spengler to be wrong. We would simply be lying to ourselves. We will only ground ourselves in truth if we acknowledge the need to engender the power that moves us forward. This, however, can only be produced out of the living spirit, out of the spirit which anthroposophically oriented spiritual science is seeking. Spiritual science has something that must enter and imbue the impulses of the present age if we are to have a life of the spirit that is Christian again, a political life that once again is human and does not fail to encompass the human being, and an economic life that is governed by human beings and not by horse-power years. Those horse-power years that human beings have alongside them are an expression of the destiny-governing principle of technological products, of something that exists outside and separate from human beings—something inhuman.

The events of recent years cannot be ascribed to human feelings and emotions. We must discern and decipher the meaning of technologically produced horse-power years, the terrible signs which Ahriman is beginning to inscribe into the evolution of humankind. A new understanding of the Christ must be found to lead us beyond this.

LECTURE 11

STUTTGART, 22 NOVEMBER 1920

LET us recall a number of things that are already quite familiar and use them as a starting point for important reflections which will, in a sense, continue the theme I discussed a few days ago.

We know that the human being is fourfold, that we may characterize ourselves as possessing a physical body, a life body, an astral or sentient body, and an I. We also know that we can only really understand human beings if we add other aspects to these four. The first four refer to aspects that are more or less fully developed at the present time, while three more must still be added—spirit-self, life-spirit, and spirit-man. These three aspects of human nature cannot be regarded as fully developed at the present time. We can merely refer to them as a potential inherent in human beings, one that will unfold in the future.

In the same way that we now have a physical body and so forth, through to the I, in time to come we shall have a spirit-self, a life-spirit and a spirit-man. We know from long-available anthroposophic literature that these different aspects or levels of the human being are connected with the whole cosmos and with cosmic evolution. In a sense we relate the physical body to the earliest embodiment of this earth, which we call Old Saturn. The life body relates to Old Sun, the astral body to Old Moon, and the principle we call our I relates essentially to the earth as it is at present.

What do we mean when we say that we relate the I we bear within us to the present earth? It means that inherent in the elements of the earth, the forces of the earth that are known to us—or perhaps unknown to us—is the principle that activates the I. Our I is intimately bound up with the forces of the earth.

If you consider the whole evolution of the human being you will find that human nature as we know it today relates largely to the past—the physical body to a far distant past, to Old Saturn, the life body to the time of Old Sun, and so forth, and that our I is not yet fully developed but in its essential nature relates to the present earth. This immediately suggests that the elements we refer to as spirit-self, life-spirit and spirit-man are not in fact rooted in the earthly realm. As human beings with the potential to evolve spirit-man, life-spirit and spirit-self, we have something in us that needs to be developed beyond this earthly realm; we will have to develop it without the earthly realm guiding us in this. As human beings we are part of this earth and our mission is in the first place to achieve full I development; to some extent we have already developed it. The forces of the earth, the intrinsic nature of the earth, served as our guide in developing the I to the extent to which we have now developed it. We shall continue with this development for the rest of earth evolution, deepening and to some extent enhancing what has developed so far, and for this we shall be indebted to the earth and its forces. Yet we also have to say to ourselves that if we were entirely dependent on the earth and its forces in developing our essential human nature, we would never be able to develop a spirit-man, a life-spirit and a spirit-self. The earth has nothing to give in this respect; it is only able to help us develop the I. In relation to human nature, therefore, the earth must be seen as something that cannot in itself make us into full human beings. We are on this earth *and* we have to go beyond it. Anthroposophic literature refers to this by showing that our evolution depends on the earth stage being succeeded by Jupiter, Venus and Vulcan periods. During those periods we will have to achieve full development, also outward development, of the spirit-self, life-spirit and spirit-man.

At present, however, we are on this earth and have to develop here. The earth cannot give us everything we need for us to progress in future to spirit-self, life-spirit and spirit-man. If we had to depend on the earth for everything we have to develop in ourselves we would have to do without spirit-self, life-spirit-and spirit-man.

It is easy to say such things in theory, but it is not enough to present them theoretically. They will only really touch us as human beings if we allow them to take hold of our whole being, if we come to feel the whole weight and burden of the riddle which lies in our having to say to ourselves: 'As human beings we are on this earth. We look around us. None of the many things the earth has to give—its beauty and its ugliness, its pain and suffering—none of the ways in which it can shape our destiny can provide what we need to become full human beings.' There must be a longing in us that goes beyond anything the earth can give. This is something we must feel, something that must bring light and warmth into all the ideals we are capable of. We must be able to ask ourselves in all seriousness and very profoundly: 'What shall we do, seeing that we have only the earth around us, and yet must progress toward something in which this earth cannot guide and serve us?' We must be able to experience, to feel, the full gravity of this question. In a sense we should already be able to say to ourselves that the earth is not enough for our needs, and that as human beings we will have to grow beyond this earthly realm.

Anthroposophy will only be able to serve human beings rightly if they are able to ask themselves such deeply felt questions; if they are aware of the gravity of such inner questions of destiny. Being aware of their gravity we can be guided in the right way to return to the Mystery of Golgotha, which has been so much a part of our last two lectures. We may be guided back to the Mystery of Golgotha to consider again the event that is to happen in this century, during the first half of the twentieth century, that will be like a spiritualized Mystery of Golgotha. In all discussion of the Mystery of Golgotha it had to be stressed that the Christ is definitely not of the earth, that the Christ entered into an earthly body from spheres beyond this earth—doing so, we can say, at exactly the right moment. In the Christ something united with this earth that came from outside it, from beyond this earth. If we really experience the Christ we are able to join our own essential nature to this super-earthly principle, and in this way gain something that will give inner strength, energy, filling us with inner warmth and light. This will take us beyond the earthly realm because it has not itself originated in that realm; because the Christ has come to earth from spheres beyond the earth.

We look with longing to the spheres beyond this earth. Longing to become full human beings—to develop the spirit-self, life-spirit and spirit-man which we shall have to develop in the future—we survey the earth and say to ourselves that the earthly realm itself does not contain what we need to develop our own nature and take it onward beyond the earth. We must turn our eyes away from the earthly realm and look to the principle that has come into the earthly realm from beyond the earth. We must look to the Christ and recognize that the Christ brought to earth the non-earthly forces that can help us to develop aspects that the earth can never help us to develop. We must take hold, with the whole of our being, of what to begin with is more in the form of concepts, of ideas. We must use this to help us recognize the Christ as the one who has come to redeem our humanity. We must come to recognize Him as the being who will enable us to overcome our union with the earthly realm, so that, as we might say, we are not buried for all eternity on earth and fail to develop our potential to evolve beyond it. When we thus come to see Christ as the one who will redeem our essential human nature, when we are able to see that there must be something within the nature of this earthly realm that will take us beyond it, when we feel that it is He who will lead us to become complete human beings—then we feel the power of Christ within us. And we really must come to realize that we cannot seriously speak of progressive development to spirit-self, life-spirit and spirit-man unless we are aware that there is no point in speaking of these things unless we appeal to the Christ, for the Christ is the principle that can take our evolution beyond anything the earth is able to give.

Basically this is also the most important issue at the present time. Many people today, particularly those in the civilized world, seek to shape terrestrial things in a certain way; they wish all human potential to be achieved through configuring earthly society itself in a particular way. That, however, can never happen. We shall never be able to evolve an entirely earthly political or economic life, nor indeed a cultural life, that would make us into full human beings. People still believe that such things are possible. They make such attempts but fail to realize that there is something in us that can only be developed further by a super-earthly principle.

Christ Jesus first appeared in a physical body at a time whose essential nature I have already described from many different points of view. We are now living in an age where He is to appear again to human beings in a form that I spoke of on the last occasion. It is clearly impossible for us today to speak of the whole scope of the renewal of the Mystery of Golgotha, but I want to refer to it again from a particular point of view.

The scientific element and everything connected with it has grown particularly strong over recent centuries, from the beginning of the fifth post-Atlantean epoch. In a recent public lecture I called it the 'science-oriented spirit of the West'.[70] This science-oriented spirit of the West did not initially relate to the Christ spirit at all. If you take an honest, unbiased look at modern science you will find that it has no real relationship to the Christ being. The best demonstration of this is the following. I have said before that Christianity first entered earth evolution at a time when remnants of ancient clairvoyance still persisted, and people grasped it with these remnants of ancient clairvoyance. Christianity then continued as a tradition, becoming increasingly diluted into mental concepts, but surviving as a tradition. Finally it became a traditional wisdom of mere phrases and words. But over the last three or four centuries the academic spirit appeared on the scene. This also addressed itself to the Gospels. A great many people did and indeed still do today revere the Gospels because they impart the secrets of Golgotha. The science-oriented spirit of the modern age, however, addressed itself to the Gospels—this was particularly in the nineteenth century—and found them to contain contradiction upon contradiction. Unable to comprehend, it interpreted the Gospels in its own way. Basically the situation is now that this academic searchlight, particularly in recent theology, has dissolved the Christ element in the Gospels. It is no longer there. If modern theologians say that the Gospels tell us something or other about the Christ they are not being entirely honest, not entirely truthful, or else they construct their theology from all kinds of conflicting ideas. So we may indeed say that modern academic thinking has destroyed the spirit of Christianity, which consisted of remnants of ancient clairvoyance and persisted as a tradition founded on those remnants.

You see, this modern spirit of enquiry was initially not imbued by the Christ spirit, and can only be so again when enlivened by spiritual vision, by what modern spiritual science is seeking to achieve.

Modern spiritual science seeks to be as scientific in its thinking as any other science. The aim is, however, not to have a dead science but to let it become inner experience, just as we have inner experience of our vital powers as human beings. This newly enlivened science will succeed in penetrating to the Christ again.

What form will this enlivened science take? Some things are already underway, but I regret to say that they have not attracted much interest. I think I ought to mention that in the early nineties—well, in fact in the late eighties—of the last century [1800s] I drew attention to a certain connection between the way in which both Schiller and Goethe developed.[78] I spoke of Schiller's attempt to solve the riddle of human evolution in his own way in his letters on aesthetic education. He started with completely abstract ideas. The first was the idea of logical necessity. He said to himself: 'This logical necessity is compelling for us human beings. We have to think logically. Freedom does not exist when logic has to be used to analyze something, for we are then subject to the laws of logic. Freedom does not exist here.' The second idea in Schiller's mind was that human beings have natural needs; this concept encompasses everything that is instinctive and arises from the human capacity for sensual desires. In this respect, too, human beings are not free but subject to necessity. In a certain way, therefore, human beings are the slaves of the highest intellectual achievement they are capable of, the logical necessity their abstract intellect is able to acknowledge by the process of reasoning. On the other hand, natural needs, human instincts, also rule and enslave human beings. It is possible, however, to find a middle position between logical thinking and instinctive feelings. Schiller felt that this middle state came to realization above all in the work of creative artists and in aesthetic pleasures. When we look at something beautiful or create something beautiful we are not thinking logically, yet we are thinking within the spiritual realm. We link ideas, but in doing so we do not pursue logical connections but surrender to aesthetic appearance. On the other hand art seeks to make everything it manifests tangible, perceptible to the senses in the same way that matters of natural necessity, of instinct, are also palpable to the senses. Schiller therefore concluded that art and aesthetic pleasures on the one hand suppress logic to a degree, so that it can no longer enslave us, so that in a sense it merges with things that we master and manage in personal terms. On the other hand art raises the instinctive element to the sphere of the spirit. In other words art enables us to feel that the instinctive element is also spiritual, and to experience logic as something personal. Schiller wanted to apply this to human life in general, by saying that in this condition human beings were not enslaved by a higher principle, nor by a lower one, but were indeed free. He wanted this to be the power that also governed society—social interplay where people met face to face. People would then find that good things were also pleasing and that they could follow their instincts because they had purified them and made them spiritual, so that these instincts could no longer drag them down. Human beings would then also share a communal life that gave rise to a free society. Thus Schiller considered three human conditions, albeit in an abstract way: the condition of ordinary physical needs, the condition of logical necessity, and the free condition of aesthetic experience.

Schiller developed this view of life in the early 1790s in his letters on aesthetic education, which he then presented to Goethe. Goethe was quite a different type of person from Schiller. He felt that Schiller was trying to solve a certain riddle, the riddle of the intrinsic nature of the human being, of human evolution and human freedom. Goethe was a more complex and profound character, however, and for him the issue could not be simply resolved by taking three abstractions and construing the whole essence of human evolution from them. Instead, his fairy tale, *The Green Snake and the Beautiful Lily*, lit up in his mind. In twenty or so different figures he embodied the human soul's potential capacities, and gave a picture of human evolution in the interplay between them. What Schiller sought to found on three abstract ideas, Goethe sought to embody in twenty imaginations. The two men understood each other in a way. What exactly was it that they had done? Schiller used a scientific approach in writing his letters on aesthetic education. He really employed the very same scientific spirit that later became that of the nineteenth century. He did not go as far as that nineteenth-century spirit of enquiry, however. He still remained, as it were, at a personal level. Nineteenth-century science completely excluded the personal aspect and took pride in being entirely impersonal. The more impersonal that knowledge can be made, the closer scientists feel they are to this ideal. Nineteenth-century scientists said, and modern scientists still say: 'We know this and that about one thing or another. We know it in a way that remains the same for every individual, and thus it contains no personal element.' Knowledge excludes the personal element to such an extent that modern people are only satisfied with their science once it has been coffined in the tombs we must regard as the mausoleums of the life of the mind and spirit today, that is, libraries. Dead knowledge is stored in libraries, and we go there when we need some bone or other that we want to include in a dissertation or in a book. Those tombs are the true ideals of the modern scientific spirit. People walk about among all the highly objective knowledge stored there, but their personal interest is somewhere else; it is definitely not contained there.

Schiller did not go as far as this in his letters on aesthetic education. He remained at the personal level. He sought personal enthusiasm, personal engagement, for every idea he developed. This is important. His letters on aesthetic education are certainly abstract, yet there is still the breath of an individual spirit in them. Knowledge was still felt to be connected with one's personal individuality. Schiller's abstract ideas therefore still contained a personal element. He did not yet allow ideas to leave that realm and enter into a completely objective and impersonal, inhuman sphere. He did, however, go as far as the development of abstract ideas. Such abstraction was impossible for Goethe. He remained with images, but he was very careful about this. He did not yet live in an age when spiritual science could be established. He felt some hesitation about sharply defining the images he presented in his *Tale of the Green Snake and the Beautiful Lily*, only hinting that he was, in fact, concerned with a future life of society. This comes clearly to expression in the conclusion of his tale. Goethe did not wish to go as far as clear, strict definitions. He did not say that society should be threefold, like the three aspects embodied in the Golden King as the king of wisdom, the Silver King as the king of outward political life, and the Bronze King embodying life in the material sphere, in the economic sphere. Goethe even represented the centralized state in the figure of the King of Mixed Metals who collapsed in a heap. He did not, however, get to the point of making sharp definitions. It was not a time when such delicate fairy-tale figures could be converted into solid characterizations of social life. I think you will agree that Goethe's figures were subtle fairy-tale figures. The time had not yet come when ideas that lived still halfway between artistic imagination and objective Imagination could be applied to outer life.

Years ago the idea of putting on a play in Munich arose, with the aim of presenting in theatrical form the world-shaping powers in Goethe's *Tale of the Green Snake and the Beautiful Lily*. This proved impossible. The whole thing had to be made much more real. The outcome of this endeavour was the Mystery Play, *The Portal of Initiation*. It is more than obvious that in Goethe's day the time had not yet come when things presented in subtle fairy-tale images could be transformed into the real characters that appear in *The Portal of Initiation*. When *The Portal of Initiation* was being written, the time had arrived to begin to carry these things into life, to realize them. It was not enough, therefore, merely to re-interpret the Golden King, the Silver King, the Bronze King and the King of Mixed Metals. We had to show that the life of society today, with its all-encompassing centralized state, must inevitably founder and collapse, and that clear distinction must be made between the life of mind and spirit (Golden King), the political element (Silver King) and the economic aspect (Bronze King). My book *Towards Social Renewal* is Goetheanistic, if properly understood, but it represents a Goetheanism of the twentieth century.

What I am saying is that Goethe and Schiller were able to reach a certain point in their day and age, Schiller in developing abstract ideas in his letters on aesthetic education, and Goethe in story images. Goethe could get very irritated when people tried to interpret his images. He had the feeling that the time had not yet come to transform these images into tangible forms applied directly to life. This shows very clearly that Schiller's and Goethe's time was not yet one when the modern scientific spirit could be allowed to become inhuman and objective; it still had to remain at a personal level. We will have to return to that level, and we can only do so with the help of spiritual science. Spiritual science must guide us to find the reality of what Schiller attempted to express in abstract ideas in his letters on aesthetic education and what Goethe, trying to solve the same riddle, hinted at in his fairy tale.

The scientific spirit has to become personal again. The earth cannot help us with this. Science itself has to become Christ-filled. By bringing the Christ idea into science we create the first beginnings of an evolution toward the spirit-self.

Let us be clear about this: the earth has prompted us to develop the I, and in its decline will prompt us to develop the I still further. This earth is something we shall have to leave behind us in order to continue our evolution on Jupiter and beyond. We cannot connect ourselves as full, complete human beings with this earth. We must in a sense reclaim our humanity from the earth. If we were to develop only the earth-related science towards which Schiller and Goethe did not want to go, Schiller by keeping his abstract ideas personal, Goethe by keeping to half-developed imaginations, and if we were only to take our promptings from earthly factors, we could never develop the spirit-self. All we could develop would be a dead science. We would therefore be adding more and more to the corpse-strewn libraries, to the piles of books, where everything human is excluded. We would walk about among these 'thought-corpses', they would cast their spell upon us, and we would thus conform to Ahriman's ideal. One of the things Ahriman wants is for us to create plentiful libraries, storing plenty of dead knowledge around us. The Ancient Egyptians walked among their tombs, even the early Christians walked about among dead bodies, and Ahriman wants us to do the same. He wants human nature to slide back more and more into mere instinct, into egotistical instincts, and he wants all the thoughts we are able to muster to be stored in libraries. It is possible to imagine that a time will come when a young gentleman or even a young lady, of twenty to twenty-three, will not initially know how to advance in the world of the Silver King—in external terms we call that taking one's PhD. Little enough surfaces from the human being's own depths. If this young person wished to write a doctoral thesis drawing on what arises from their own human nature—I am of course assuming that a time may come when Ahriman has won the day!—such a thesis would be rejected as being subjective and personal. The young person would therefore visit libraries, reading one book after another, preferably indexes of references relating to a particular keyword. A new keyword would mean taking out yet another book. The whole thing would then be collated to create a thesis. Only their outer, physical individual would actually be involved in all this, however. The young man or woman would be sitting at a desk piled with books. Personal involvement would consist in getting hungry when one has been at it for a few hours, and this hunger would be felt to affect one personally. Personal involvement might also arise through having human relationships, with certain commitments that would have to be acknowledged when they came to mind after a few hours in the library. At that point the student would close the books, and any personal connection with them would cease. The thesis developed from what one has found in various books would go into yet another book, a small one or a large tome; it would go to join the others on the library shelves and wait for someone else to come and use it. I am not sure if we have already reached this stage or not today, but if Ahriman's ideal ever comes to realization that is exactly how it will be. It would be a terrible situation. Human individuality would wither away in such a terrible objective, non-human and impersonal situation.

To combat this, knowledge has to become a personal matter. Libraries should shrink if possible, and people should carry the things that are written in books in their souls. Spirit-self can only develop out of knowledge made personal, and that cannot happen unless people learn about the things that are not of this earth. The earth has passed the mid-point of its evolution. It is dying. Knowledge is dying in our libraries. It is also dying in our books, for they are the coffins of knowledge. We must take this element of knowledge back into our individuality. We must carry it in us. Help will come above all from the renewal of the Mystery of Golgotha. This will help people with knowledge; it will help the followers of the Golden King.

New life must also come in another sphere, the sphere of rights. Human beings have as little personal connection with the legal system nowadays as they have with the sphere of knowledge. I presented a small but definite proof of this in a recent public lecture.[79] I said that the German Empire had free and equal general suffrage. You could not have asked for anything better. But did those voting rights relate to life? Did people cast their votes in a way that accorded with this franchise? Did something come to life in the configuration of the German Empire because of this franchise? Absolutely not. The franchise was merely inscribed in the Constitution. It was not alive in people's hearts. A time must come when people will no longer need to enshrine in an objective Constitution how one human being should relate to another; then, instead, living relationships between people will give rise to law that is also alive. What need is there for written constitutions when people have the right feeling for their relationship as one human being to another and when this relationship comes to be a personal matter? In the last three decades of the nineteenth century, human relations grew impersonal, and they have remained impersonal under the powerful materialism of the twentieth century. The law will only come alive when human beings have the Christ spirit within them.

In the sphere of rights, then, people must become followers of the Silver King. In economic life, on the other hand, they must become followers of the Bronze King. This means no more and no less than that the abstract ideal of brotherhood or companionship must become real. How can fraternity become real? By associating, by truly uniting with the other person, by no longer fighting people with different interests but instead combining those different interests. Associations are the living embodiment of brotherhood. The life-spirit must be alive in the sphere of rights, and with the Christ spirit introduced into economic life, the germinal beginnings of spirit-man will come to life through associations. The earth itself, however, yields none of this. Human beings will only come to this if they let the Christ, who is now approaching and appearing to us etherically, penetrate them fully.

You see, therefore, that what we can call the spiritual renewal of the Mystery of Golgotha relates to what anthroposophic cosmology teaches. We come to see this when we can say to ourselves that we have the potential to develop spirit-self, life-spirit and spirit-man. Our thinking has grown so abstract, however, that it seems terribly dry and prosaic to hear that something as sublime and spiritual as spirit-man must first of all manifest in associations created in economic life—in that 'low' economic life concerned with material things. Surely a spiritual scientist cannot refer to economic life without 'lowering' himself? A spiritual scientist must supposedly convene with people in circles where no one speaks of anything as mundane as food and drink, and where one lives entirely in 'the spirit', which in fact means in abstract ideas.

The fact is, however, that when these people have been sitting in their higher circles or sects for long enough and have found their inner gratification there, they will finally emerge and of course take bread and—well, let us say 'water', lest we really offend. As a rule, terribly little of all the principles they have established to gratify their souls in their conventions will find application in actual life.

The true life of the spirit exists only where it is strong enough to penetrate material life—and not leave it to one side as something that enslaves and compels us. This is something you really need to acknowledge.

I think when we come to consider things like these we realize the seriousness needed in our approach to life today. Yet this seriousness can only develop if we enter into things as deeply as spiritual science enables us to do. You see, the spiritual can only be brought close to human individuals through spiritual science. In a way Schiller and Goethe were the last who could still remain at the personal level, and this was due to something still accessible to them from the past. Schiller did not allow abstract ideas to develop the icy coldness of modern ideas. Goethe kept his imaginations at a personal level and did not let them break through fully into life's outward reality.

Today we must go beyond this point. In the rough and tumble of modern reality we cannot get anywhere much with aesthetic letters—except maybe at aesthetic tea parties—nor with 'fairy tales'. At most one might perhaps have beautiful conversations about them in salons and in those caricatures of salons, the new-fangled lecture theatres where literary history is taught in the modern way. What is needed today is that we break through into life with the things that Goethe and Schiller still kept at the personal level. This will need powerful ideas and on the other hand also powerful imaginations: a true spiritual understanding of the outer world must arise. To achieve this, we must fill ourselves with the Christ spirit. We all need belief in the Christ spirit in its true sense, belief that the Christ principle is something we have to unite with the part in us, as human beings, that will take us beyond this earth and make us full human beings by helping us to develop spirit-self, life-spirit and spirit-man.

All the things we encounter in spiritual science have an inner connection with each other. Understanding these inner connections we shall be able to see spiritual science in the right light and know that it belongs to the present era. We shall also know that in the present age spiritual science must gain a very real influence in all spheres of practical life.

This means, however, that spiritual science must take the whole of life extremely seriously. A true spiritual scientist would feel the lack of real seriousness to be frivolous, a failure to do more than fashion beautiful abstract ideas that may gratify the soul but are unable to break through into actual life.

This is something which has been weighing heavily on spiritual science for more than a year; it has been weighing heavily on those of us working here in Stuttgart. This work in Stuttgart has now presented us with the responsibility of bringing spiritual science to bear on the practical life that surrounds us everywhere. Principles that Goethe presented in fairy-tale images of a Golden, a Silver and a Bronze King, and a King of Mixed Metals who collapsed in a heap, must now be brought to bear on life and must become the threefold social order. You will remember that when the King of

Mixed Metals collapsed in a heap, certain persons came and licked up all the gold. If you take a good look at the world around us today you will see this phenomenon. In November 1918, Central Europe's King of Mixed Metals collapsed, and don't you see now how the various ministers who have held office since that time, the various leaders, are licking away and will go on licking until they have sucked out all the gold? Then the whole form of the Mixed King, a form empty of all spirit, will collapse, and people will be horrified. So we really ought to be serious—not about fairy-tale images of a Golden, a Silver, and a Bronze King, but with firm understanding of the three elements of the social organism: the cultural and spiritual sphere, the political sphere, that is, the state, and the economic sphere.

It has to be said, however, that when one comes to speak of these things, two thoughts immediately come to mind. One of these I want to talk about today, for the longer we have to go on working like this in Stuttgart the more obvious it becomes that, for the time being at least, it is simply impossible to find time to talk to the friends who have been used to coming and asking my advice in earlier years. For a long time now I have had to put people off when they wanted to discuss things that previously could be discussed in private, postponing these discussions for a later date. Although my visits have been getting longer and longer, all efforts have had to be concentrated on the great task in hand. I feel it really has to be said that, this time in particular, it has been quite impossible to consider personal requests. This is as painful for me as it is for you and I know that we cannot go on like this in the long run, for that would deprive the anthroposophic movement of its foundations. In that case we would be building on shifting sands.

On the other hand it also has to be realized that people always like to cling to the old ways. Yet we are doing something entirely new in really getting to grips with what can be called the Golden, the Silver and the Bronze King. This is an extremely serious matter. Spiritual science cannot do such a thing as licking the gold away from the King of Mixed Metals who collapses in a heap; and some people take this amiss. I know I am poking around in a hornets' nest, but

I shall have to poke around in quite a few hornets' nests, for example by characterizing a person such as Hermann Keyserling[80] as someone who is simply not telling the truth, who is telling lies.

Some people say there is too much criticism issuing from the anthroposophic movement today. But let me repeat once again what I have said many times before: these people see what we have to do in order to defend ourselves—and they take exception to this. Exception is even taken by people who are sitting in this room and listening to the things that are being said. And they never repudiate those who throw dirt at us from outside—for that would mean becoming argumentative oneself. It is considered unkind for an anthroposophist to call someone a liar, when that is in fact the truth. Yet anyone who wants to tell lies about the anthroposophic movement is allowed to fling any kind of lie at us. The journal of our movement for a threefold order is often considered too polemical. But let us stand up to those whom we necessarily have to dispute with; let us have the courage to address our words to them and not against ourselves, for we are simply forced to defend ourselves. An old, bad habit is that people are more interested in an anthroposophy that provides self-gratification than in a serious anthroposophy which considers the great problems of the present.

Now and then it is really necessary to speak very seriously about these things. The things I said about Count Hermann Keyserling in my public lecture, for instance, relate not only to the attacks on anthroposophy from that quarter; they relate to the whole inner insincerity of that kind of intellectual life. Read the chapter entitled 'What We Need. What I Want' in his most recent book.[81] It does not refer to anthroposophy, but you will find there the whole empty, insubstantial conceptual schematism which empty-headed people think is so useful. That of course is the great evil in our time, that people reject the things that take their substance from the spirit—the living spirit—and want only to have empty words, mere shells of words.

If people go on wanting things like this they will destroy humanity. The hollow phrases coming from that source—even if they are called the *Diary of a Philosopher*[82]—undermine the whole of human culture. What are they, these hollow phrases? They are the phrases one produces if one licks the King of Mixed Metals. You may be fairly brutal in your licking, like some of the socialist leaders today, or you may be wearing elegant patent-leather boots like Count Keyserling—it really makes no difference.

I may be putting these things sharply, but please do not think this reflects an emotional involvement. They are put sharply because it has to be said, unfortunately, that there are some who wish to be counted among anthroposophists but whose hearts are not really in it. They cannot be sufficiently serious; they do not wish to be sufficiently serious, they do not want their hearts to be involved. It is not being unkind to speak the truth when it is necessary to do so. But let me ask you if it is kind of anyone, who wants to be one of us, to allow others to sling mud at us and then call us unkind when we have to defend ourselves? It may seem regrettable that we have to use sharp words to defend ourselves, but precisely for this reason we should keep words sharp, not indulge in wishy-washy feelings and the like, and merely start repeating the rubbish that polemics are always unkind and unjustifiable.

The difficulty is that within the movement that is to develop as the anthroposophic movement we find so few people who are wholeheartedly with us. To achieve the kind of thing that we are meant to achieve through the anthroposophic movement, we need many such individuals today. We have found dedicated people in many different fields, above all in education, the Waldorf School teachers. We have also found dedicated individuals in other fields—but it is simply not enough. The number of those who simply do not want to become fully involved is extremely large, right here in our own ranks, and yet we need people to be fully dedicated to our cause. That is why we are making so little progress. As time has gone on we have found again and again that, when it really comes down to it, many of the people who have joined us to hear the things said within the movement are somehow embarrassed to declare themselves openly for us in the outer world. We have heard it said again and again that it would be better not to use the name 'anthroposophy' in public; that one should leave the name out and 'slip things in here and there' that refer to anthroposophy. That is the delightful way people who do not want to take anthroposophy seriously like to put it. So the gentleman, or particularly the lady, intends to 'slip something in' here and there by way of anthroposophy, because she or he feels ashamed to speak openly about it. So they 'slip things in'! You won't have to be all that courageous, then, and you won't create any awkwardness—'just let it slip in'.

Now is not the time to let things slip in, however. It is time to be open and honest and to use words that tell the truth about things. The people who are against us do not let things slip in, they put things bluntly. And it should be considered an outrage by all who have joined our ranks that someone like Hermann Keyserling has the cheek to say that this spiritual science of ours materializes the life of the spirit, that it is a physical science of the spirit. We know that this man used sneaky ways to get hold of our lecture courses from a large number of people, in order to find out what is said in them; and all one can say is that in writing the things he is writing today he is quite deliberately writing untruths. We call it lying. Anyone who objects to our saying this is a lover of lies. Anyone who says that we are too argumentative when we are rightly speaking the truth has no feeling for the truth and is a lover of lies. The love of lies should not be our business in the anthroposophic movement, for we must love the truth. You must feel the whole weight of these words: to love the truth; not to love lies for the sake of convention, for the sake of a pleasant social life. To be easy-going when it comes to lies is just as bad as loving them. In the immediate future the world will not progress through frivolous indifference where lies are concerned, but only if we freely and openly profess the truth. Anthroposophy has to consider serious and sublime spiritual matters, and we have never failed in this. Anyone who says that it is spiritual materialism to speak of Saturn, Sun and Moon when they are free to open my *Occult Science* and read what it says about Saturn, Sun and Moon there, is lying indeed. The book says nothing about making the spirit something material. People cannot be aware of the true seriousness of the situation if they ask that we use polite, untruthful terms to address mud-slinging opponents.

Such things in fact are intrinsic to real love. Real love demands enthusiasm for the truth. The world will only progress if we show enthusiasm for the truth.

There are profound spiritual reasons why I have to say these things today, as I am about to leave you again for a while. I am very sorry that I am quite unable to talk to individuals at present, because there simply is not the time. Yesterday the friends of our movement for a threefold order, and of the *Kommende Tag*[39] were again in session until 3 o'clock in the morning, and that is how it continues, more or less every day. I regret that many things have to be left aside, things that people have come to love. On the other hand there may be hope after all that, in view of the widespread efforts now being made, the anthroposophic movement will gain the rightful place in this world that it must gain, because it has the strength and the will to progress by speaking the truth. If we are to work in the truth, then we can do nothing other today than show untruthfulness in its true light when it becomes as blatant as it has.

I needed to remind you of our commitment to the truth. It is most necessary for all of us, dear friends, to let this longing for the truth fill all of us, each and every one. If it is still within the bounds of human capacities, then this spirit in which we long for the truth will be the only thing that can prevent the barbarism that otherwise must come upon the human race. It will be the only spirit in which we shall progress toward a new culture of the spirit.

Notes

Sources. The lectures were taken down by a number of stenographers whose names are not recorded. The original German text of the book was based on the full texts subsequently transcribed by the stenographers, which are housed in the archives of the Rudolf Steiner Estate Administration in Dornach.

The German title of the book was not given by Rudolf Steiner but by the editors.

Other works by Rudolf Steiner that form part of the German Collected Works (*Gesamtausgabe*, GA) are referred to by their GA numbers in these notes.

1 Ludwig Andreas Feuerbach, German philosopher. Referring to Feuerbach's anthropomorphism, Rudolf Steiner repeatedly quoted from his book *Das Wesen des Christentums* (1841), e.g. in a lecture given in Dornach on 15 October 1921 (in GA 207, *Cosmosophy*, Vol. 1).
Ludwig Buechner, German physician and materialist philosopher.

2 'Die Weisheitslehren des Christentums im Lichte der Theosophie', Colmar, 21 November 1905 (no transcript extant).

3 Details not known.

4 Paul Deussen, philosopher and Sanskrit scholar.

5 Richard von Garbe, Sanskrit scholar.

6 Johann Gottlieb Fichte, German philosopher. *Die Bestimmung des Menschen*, 1800.
Friedrich Wilhem Joseph von Schelling, German philosopher. *Bruno, oder über das göttliche and natürliche Princip der Dinge. Ein Gespräch.*
Johann Christoph Friedrich von Schiller, German dramatist, poet and historian. *Briefe über die ästhetische Erziehung des Menschen*, 1795.

7 Ralph Waldo Trine, American writer.

8 Stuttgart, 21 December 1919. In *Weltsilvester and Neujahrsgedanken,* GA 195 (*Cosmic New Year*).

9 Adolphe Ferrière. *La loi du progres economique et la justice sociale. Suisse-Belgique Outremer,* July-August 1919; 1: no. 3-4.

10 Grigory Efimovich Rasputin, reputed 'holy man' who influenced the Tsarina and Tsar Nicholas II of Russia.

William II (1859–1941), third German emperor and ninth King of Prussia.

11 Dr Roman Boos, anthroposophic lecturer, writer in the field of social sciences. Pioneer of the threefold movement.

12 *Plaisanterie*—joke; *méchanceté*—spitefulness, malice.

13 Prince-Archbishop Johann Baptist Katschthaler, Salzburg, Austria. Pastoral letter of 2 February 1905 entitled 'Die dem katholischen Priester gebührende Ehre' ('The honour due to Catholic priests'), reprinted in Carl Mirbt, *Quellen zur Geschichte des Papstums and des römischen Katholizismus,* 5th edition, Tübingen 1934, p. 497 ff. The passage relating to the powers of consecration is given below.

Honour your priests, for they have the power of consecration. Catholic priests have this wonderful power of consecration, Protestant pastors do not. This power of consecration, to make the Body of the Lord be present, with the precious Blood, with the whole of His sacred humanity and His divine nature in the Bread and the Wine—that is a great and sublime power, a truly extraordinary power! Where in heaven is a power like that of the Catholic priest to be found? Among the angels? Or does the Mother of God have it? Mary conceived Christ, the Son of God, in her womb and gave birth to Him in the stable in Bethlehem. Yes. But consider what happens during Holy Mass! Does not the same thing happen, as it were, when the priest raises his hands in blessing during the consecration? Christ is really and truly made to be present, to be reborn, as it were, in the Bread and the Wine. Mary bore her child in Bethlehem, wrapping it in swaddling clothes; a priest does the same, as it were, placing the wafer on the corporal cloth. Mary brought her child into the world just once. But see, a priest does this not once, but many hundreds and thousands of times, each time he celebrates the Mass. There, in the stable, the child which Mary gave to the world was small, capable of suffering and mortal. Here, on the altar, in the hands of the priest, we have Christ in His glory, not capable of suffering and also immortal, sitting in heaven to the right of the Father, glorious and triumphant, perfect in every regard. Do priests merely make the Body and Blood be present? No. They sacrifice, offering sacrifice to the Heavenly Father. This is the same sacrifice that Christ brought by shedding His blood on Calvary and bloodlessly at the Last Supper. There the eternal High Priest Jesus Christ sacrificed His Flesh, His Blood and His very Life to the Heavenly Father; here, at the Mass, He does the same through his representative, the Catholic priest. He ordained priests to take His place so that they might continue the same sacrifice that He had brought. He has transferred to them the authority over His sacred humanity, giving them power over His body, as it were. A Catholic priest is able not only to make Him be present on the altar, lock Him up in the tabernacle, take Him out again and give Him to the faithful to eat, he is actually able to offer Him, the Son of God become Man, as a bloodless sacrifice for the living and the dead. Christ, the only begotten Son of the Father by whom heaven and earth were created and who sustains the whole world, submits to the Catholic priests in this respect.

14 'Suprema lex regis voluntas'. William II (see note 10) wrote this in the *Golden Book of the City of Munich*. See also J. von Kürenberg, *War alles falsch? Das Leben Kaiser Wilhelms II.*, Basle-Olten 1940, p. 190.

15 Dionysius the Areopagite (an Athenian converted by St Paul): 'The Heavenly and Ecclesiastical Hierarchies' and 'The Hierarchy of the Church', in *Die angeblichen Schriften des Areopagiten Dionysius, übersetzt and mit Abhandlungen begleitet von J.G. V. Engelhardt*, part 2, Sulzbach 1823.

16 Aurelius Augustinus, *De Civitate Dei.* libri XII.
Dante Alighieri, *De Monarchia.*

17 Wycliffe, John, English reformer.
Hus, or Huss, John, Bohemian reformer, burned when the Council of Constance condemned his and Wycliffe's writings.

18 Luke 17:20-21. Rudolf Steiner was quoting from the Luther Bible, but saying 'gods' instead of 'God'.

19 Grimm, Herman, *Fragmente* (vol. 1), Berlin & Stuttgart, p. 212: 'For the people of today [1891], it is no longer the Wars of Liberation fought against Napoleon I that are the last important historical event in their lifetime, but the Wars of Liberation fought against Austria and France in the 1860s and 1870s. We have been a nation where children had to be taught that they would never be permitted to act freely in influencing the

destinies of their country. Today, however, Germans are compelled to act in this way. Fifty years ago it would have been unthinkable to make it part of education to tell children that they would one day be citizens of a great united German Empire and that their duties towards God, the Emperor and their country would one day also consist in having to elect someone to represent them in a German Parliament, basing their choice on their personal judgement of the country's needs. Just to utter such things would have sounded like high treason and might have ruined the life of the person who uttered them.'

20 'Der Weg zum gesunden Denken und die Lebenslage des Gegenwartsmenschen', Stuttgart 8 June 1920. Published in *Geisteswissenschaft und die Lebensforderungen der Gegenwart,* no. 6, Dornach 1950. Reprinted in GA 335 (no English translation).

21 Lenin (formerly Ulyanov), Vladimir Ilyich, Russian revolutionary.
Trotsky, Leon (Lev Davidovich Bronstein), Russian revolutionary leader in 1917.

22 *Basler Vorwärts*, 2 June 1920, 'Die Politik der Sowjetregierung auf dem Gebiete der Religion'.

23 See Franz Brentano. *Die Psychologie des Aristoteles*, Mainz 1867. S. 199 ff.

24 Traub, Friedrich, professor at Tuebingen University, author of 'Rudolf Steiner as a philosopher and theosophist', Tuebingen 1919.

25 For details relating to this see Boos, Roman, 'Aktenmäßige Darstellung der Hetze gegen das Goetheanum' in Rudolf Steiner/Roman Boos: *Die Hetze gegen das Goetheanum*, Arlesheim 1920.

26 Refers to the Council's rejection of the 'trichotomy'. Rudolf Steiner referred to this on a number of occasions, e.g. in *Building Stones for an Understanding of the Mystery of Golgotha*, GA 175, Lectures 1 and 2.

27 'Mathilde Reichardt, a lady who published a book on science and moral philosophy in the form of letters to Moleschott in 1856, is able to lay undoubted and unenviable claim to rank first among those who turn moral concepts upside down. It is the opinion of this lady that moral philosophy merely has to ask whether a human individual develops his or her inherent traits in a balanced way. Yet nature—so she says—expresses a different intent in every human individual. She therefore does not hesitate to state "that when someone has an inherent tendency to cheat and to steal, such a person can only be an entirely moral person if he or she is a cheat or a thief." "Someone

born to be a thief has also been born with the right to live out those inherent tendencies, developing them to the full, for that is the only way in which he or she can be a strong and moral individual. And what applies to thieves also applies to all other vices, including those born to be murderers." '(Cited in Jurgen Bona Meyer, *Philosophische Zeitfragen*, Bonn 1874, p. 323 f.)

28 Vogt, Carl, champion of Darwinism.
Moleschott, Jakob, physiologist.
Buechner, Ludwig, German physician and materialist philosopher.

29 *The Philosophy of Freedom* (1894), GA 4.

30 'Die Erziehung und der Unterricht gegenüber der Weltlage der Gegenwart', Stuttgart 10 June 1920. Published in *Geisteswissenschaft und die Lebensforderungen der Gegenwart 6*, Dornach 1950. Reprinted in GA 335 (no English translation).

31 *Towards Social Renewal* (1919), GA 23.

32 See Rudolf Steiner's *Autobiography*, GA 28.

33 *Riddles of the Soul*, 1917, GA 21.

34 Molt, Emil, see ref. 37.
Kühn, Hans, writer and publisher.
Unger, Carl, engineer, owner of a machine-tool factory, member of the Council of the German Anthroposophical Society from 1905, lecturer and writer. Shot by a mentally ill individual in Nuremberg in 1929.
Leinhas, Emil, businessman, managing director of *Der Kommende Tag* AG (see ref. 39), writer.

35 Seiling, Max, first a follower then an opponent of Rudolf Steiner. Bore the title Privy Councillor.
Rudolf Steiner used word-play here, calling Hofrat (Privy Councillor) Seiling an Un-Rat. The word Unrat means garbage, refuse, ordure.

36 'Appeal to the German nation and the civilized world', pamphlet, 1919, reprinted in *Towards Social Renewal* (ref. 31).

37 Appeal to establish a Council for Culture, Whitsun 1919, signed by many well-known figures in cultural life.

38 Waldorf School: integrated primary and secondary school established by Emil Molt in 1919 for the children of workers in the Waldorf-Astoria Cigarette Factory and the general public. This was done under the guidance of Rudolf Steiner who also appointed the teachers and himself gave the preparatory teachers' seminars.

[39] *Der Kommende Tag*, joint-stock company for the promotion of economic and cultural values, established in Stuttgart in March 1920. Rudolf Steiner was Chairman of the Board until 1923. The company became a victim of post-war inflation in Germany and had to be dissolved in 1925.

[40] Stuttgart, 16 June 1920; published in *Landwirtschaft und Industrie/Neuordnung des Bodenrechtes als soziale Forderung der Gegenwart*, reprinted in GA 335 (no English translation).

[41] Spengler, Oswald, German historian. *Der Untergang des Abendlandes* (*Decline of the West*, C.F. Atkinson tr.) Munich, vol. 1, 1918, vol. 2, 1922.

[42] *Die Philosophie des Thomas von Aquino.* Three lectures given in Dornach on 22-24 May 1920. GA 74 (*The Redemption of Thinking*).

[43] *Durch den Geist zur Wirklichkeits-Erkenntnis der Menschenrätsel: Philosophie und Anthroposophie. Vier Märchen (aus den Mysteriendramen). Anthroposophischer Seelenkalender. Der Seelen Erwachen, 7. und 8. Bild.*, Berlin 1918.

[44] Eckhart, Johannes, known as Meister Eckhart, German mystic.

[45] *Knowledge of the Higher Worlds*, GA 10.

[46] The Group: a sculpture in wood at the Goetheanum in Dornach, Switzerland, which depicts the Representative of Humanity between Lucifer and Ahriman.

[47] Dessoir, Max. *Vom Jenseits der Seele. Die Geheimwissenschaften in kritischer Betrachtung*, p. 254 ff. Stuttgart 1917.

[48] *Die geistige Führung des Menschen und der Menschheit*, 1910, GA 15.

[49] Prof. Hugo Fuchs, Goettingen. See also Rudolf Steiner, 'Ein paar Worte zum Fuchs-Angriff' ('A few words on the attack made by Fuchs'), in *Dreigliederung des sozialen Organismus* 2nd year, no. 5 (Aug. 1920); reprinted in *Aufsätze zur Dreigliederung des sozialen Organismus und zur Zeitlage* 1915-1921, GA 24 (*Renewal of the Social Organism*).

[50] *Dreigliederung des sozialen Organismus* 2nd year, no. 4, supplement (July 1920).

[51] Atwood, George, English mathematician, invented a machine to illustrate the motion of a body falling under the action of gravity.

[52] *From Jesus to Christ*, GA 131.

[53] *Occult Science, an Outline*, 1910. GA 13.

[54] Wasmann, Erich, entomologist. Major researcher on ants. The MS of the lecture is not quite clear at this point, and the German editors have added the words in square brackets.

[55] Goethe said the following: 'It is said that the truth lies halfway between two contradictory opinions. Far from it! The problem lies in between,

the imperceptible, ever-active life conceived in tranquillity.' *Maxims und Reflections.*

The MS of the lecture was quoted verbatim in the earlier German edition but has been found to be incomplete and therefore was revised for the German edition on which this translation is based.

56 Christlieb, Max, a friend from Rudolf Steiner's time at Weimar who had done much to make *The Philosophy of Freedom* more widely known when it had just been published. See also *Rudolf Steiner and Christlieb. Ludwig Kleeberg, Wege und Worte*, 2nd edition 1961 p. 76 ff, where reference is also made to the encounter in Marburg.

57 'Die grossen Aufgaben von heute im Geistesleben, Rechtsleben und Wirtschaftsleben. Eine dritte Gegenwartsrede', in Rudolf Steiner: *Drei Gegenwartsreden*, Dornach 1952. Reprinted in GA 335 (no English translation).

58 See conclusions of lectures given on 11 and 17 January 1920 in *Geistige und soziale Wandlungen in der Menschheitsentwickelung*. GA 196, p. 52 & 82 ff. (See *What is Necessary in These Urgent Times.*)

59 Rohm, Karl, editor of the journal *Der Leuchtturm*, Lorch (Württemberg).

60 Further details in E. Uehli, 'Die gestohlene Dreigliederung' in *Dreigliederung des sozialen Organismus* 2nd year, no. 11 (Sept. 1920).

61 Knapp, Alfred, established the Internationaler Orden für Ethik und Kultur ('International Order for Ethics and Culture').

62 Tagore, Sir Rabindranath, *Nationalism*, Penguin 2010.

63 James, William, American psychologist and pragmatic philosopher, author of *The Varieties of Religious Experience.*

64 'The Tale of the Green Snake and the Beautiful Lily' in *Unterhaltungen deutscher Ausgewanderten* 1795, p. 225 ff.

65 Humboldt, Karl Wilhelm von, *Ideen zu einen Versuch, die Grenzen der Wirksamkeit des Staates zu bestimmen*, 1792.

66 Wilson, Thomas Woodrow, *The State. Elements of Historic and Practical Politics.*

67 Keely, John Worrell. See also lecture given by Rudolf Steiner in Dornach on 1 Dec. 1918 in GA 186.

68 The Franco-Prussian War 1870–71.

69 Throughout this lecture Rudolf Steiner used an invented word, *Technizismen* (technicisms) for 'products of technology'.

70 Stuttgart, 10 November 1920, 'Die Geisteskrisis der Gegenwart und die Kräfte zum Menschheitsfortschritt', published in GA 335 (no English translation).

71 Bayle, Pierre, French philosopher. The statement has not been traced.

72 Goethe, Johann Wolfgang von, *The Sorrows of Werther*, 1775.

73 Miller, Johann Martin. *Siegwart*, a novel published in 1776 (2 years after Werther), very popular for a time.

74 Paquet, Alfons, writer. *Im kommunistischen Russland* appeared in 1919.

75 See also public lecture referred to in note 69.

76 Clemenceau, Georges. French physician and politician.

77 Harnack, Adolf von. German theologian.

78 'Goethe als Vater einer neuen Aesthetik', lecture to the Goethe Society in Vienna on 9 November 1888; first published in Vienna in 1889, reprinted in *Methodische Grundlagen der Anthroposophie 1884–1901,* GA 30 and *Kunst und Kunsterkenntnis,* GA 271 (*Goethe as the Founder of a New Science of Aesthetics*, Anthroposophical Publishing Co., London 1922).

79 Stuttgart, 16 November 1920, 'Die Wahrheit der Geisteswissenschaft und die praktischen Lebensforderungen der Gegenwart', published in GA 335 (no English translation).

80 Keyserling, Hermann, Graf., German philosopher.

81 Keyserling, Hermann, *Philosophie als Kunst*, Darmstadt 1920.

82 Keyserling, Hermann, *Das Reisetagebuch eines Philosophen*, Darmstadt 1919.

Rudolf Steiner's Collected Works

The German Edition of Rudolf Steiner's Collected Works (the *Gesamtausgabe* [GA] published by Rudolf Steiner Verlag, Dornach, Switzerland) presently runs to 354 titles, organized either by type of work (written or spoken), chronology, audience (public or other), or subject (education, art, etc.). For ease of comparison, the Collected Works in English [CW] follows the German organization exactly. A complete listing of the CWs follows with literal translations of the German titles. Other than in the case of the books published in his lifetime, titles were rarely given by Rudolf Steiner himself, and were often provided by the editors of the German editions. The titles in English are not necessarily the same as the German; and, indeed, over the past 75 years have frequently been different, with the same book sometimes appearing under different titles.

For ease of identification and to avoid confusion, we suggest that readers looking for a title should do so by CW number. Because the work of creating the Collected Works of Rudolf Steiner is an ongoing process, with new titles being published every year, we have not indicated in this listing which books are presently available. To find out what titles in the Collected Works are currently in print, please check our website at www.rudolfsteinerpress.com (or www.steinerbooks.org for US readers).

Written Work

CW 1 Goethe: Natural-Scientific Writings, Introduction, with Footnotes and Explanations in the text by Rudolf Steiner
CW 2 Outlines of an Epistemology of the Goethean World View, with Special Consideration of Schiller
CW 3 Truth and Science
CW 4 The Philosophy of Freedom
CW 4a Documents to 'The Philosophy of Freedom'
CW 5 Friedrich Nietzsche, A Fighter against His Time

CW 6	Goethe's Worldview
CW 6a	Now in CW 30
CW 7	Mysticism at the Dawn of Modern Spiritual Life and Its Relationship with Modern Worldviews
CW 8	Christianity as Mystical Fact and the Mysteries of Antiquity
CW 9	Theosophy: An Introduction into Supersensible World Knowledge and Human Purpose
CW 10	How Does One Attain Knowledge of Higher Worlds?
CW 11	From the Akasha-Chronicle
CW 12	Levels of Higher Knowledge
CW 13	Occult Science in Outline
CW 14	Four Mystery Dramas
CW 15	The Spiritual Guidance of the Individual and Humanity
CW 16	A Way to Human Self-Knowledge: Eight Meditations
CW 17	The Threshold of the Spiritual World. Aphoristic Comments
CW 18	The Riddles of Philosophy in Their History, Presented as an Outline
CW 19	Contained in CW 24
CW 20	The Riddles of the Human Being: Articulated and Unarticulated in the Thinking, Views and Opinions of a Series of German and Austrian Personalities
CW 21	The Riddles of the Soul
CW 22	Goethe's Spiritual Nature and its Revelation in 'Faust' and through the 'Fairy Tale of the Snake and the Lily'
CW 23	The Central Points of the Social Question in the Necessities of Life in the Present and the Future
CW 24	Essays Concerning the Threefold Division of the Social Organism and the Period 1915-1921
CW 25	Cosmology, Religion and Philosophy
CW 26	Anthroposophical Leading Thoughts
CW 27	Fundamentals for Expansion of the Art of Healing according to Spiritual-Scientific Insights
CW28	The Course of My Life
CW 29	Collected Essays on Dramaturgy, 1889-1900
CW 30	Methodical Foundations of Anthroposophy: Collected Essays on Philosophy, Natural Science, Aesthetics and Psychology, 1884-1901
CW 31	Collected Essays on Culture and Current Events, 1887-1901
CW 32	Collected Essays on Literature, 1884-1902
CW 33	Biographies and Biographical Sketches, 1894-1905
CW 34	Lucifer-Gnosis: Foundational Essays on Anthroposophy and Reports from the Periodicals 'Lucifer' and 'Lucifer-Gnosis,' 1903-1908
CW 35	Philosophy and Anthroposophy: Collected Essays, 1904-1923
CW 36	The Goetheanum-Idea in the Middle of the Cultural Crisis of the Present: Collected Essays from the Periodical 'Das Goetheanum,' 1921-1925

CW 37 Now in CWs 260a and 251
CW 38 Letters, Vol. 1: 1881-1890
CW 39 Letters, Vol. 2: 1890-1925
CW 40 Truth-Wrought Words
CW 40a Sayings, Poems and Mantras; Supplementary Volume
CW 42 Now in CWs 264-266
CW 43 Stage Adaptations
CW 44 On the Four Mystery Dramas. Sketches, Fragments and Paralipomena on the Four Mystery Dramas
CW 45 Anthroposophy: A Fragment from the Year 1910

Public Lectures

CW 51 On Philosophy, History and Literature
CW 52 Spiritual Teachings Concerning the Soul and Observation of the World
CW 53 The Origin and Goal of the Human Being
CW 54 The Riddles of the World and Anthroposophy
CW 55 Knowledge of the Supersensible in Our Times and Its Meaning for Life Today
CW 56 Knowledge of the Soul and of the Spirit
CW 57 Where and How Does One Find the Spirit?
CW 58 The Metamorphoses of the Soul Life. Paths of Soul Experiences: Part One
CW 59 The Metamorphoses of the Soul Life. Paths of Soul Experiences: Part Two
CW 60 The Answers of Spiritual Science to the Biggest Questions of Existence
CW 61 Human History in the Light of Spiritual Research
CW 62 Results of Spiritual Research
CW 63 Spiritual Science as a Treasure for Life
CW 64 Out of Destiny-Burdened Times
CW 65 Out of Central European Spiritual Life
CW 66 Spirit and Matter, Life and Death
CW 67 The Eternal in the Human Soul. Immortality and Freedom
CW 68 Public lectures in various cities, 1906-1918
CW 69 Public lectures in various cities, 1906-1918
CW 70 Public lectures in various cities, 1906-1918
CW 71 Public lectures in various cities, 1906-1918
CW 72 Freedom—Immortality—Social Life
CW 73 The Supplementing of the Modern Sciences through Anthroposophy
CW 73a Specialized Fields of Knowledge and Anthroposophy
CW 74 The Philosophy of Thomas Aquinas
CW 75 Public lectures in various cities, 1906-1918
CW 76 The Fructifying Effect of Anthroposophy on Specialized Fields
CW 77a The Task of Anthroposophy in Relation to Science and Life: The Darmstadt College Course
CW 77b Art and Anthroposophy. The Goetheanum-Impulse

CW 78 Anthroposophy, Its Roots of Knowledge and Fruits for Life
CW 79 The Reality of the Higher Worlds
CW 80 Public lectures in various cities, 1922
CW 81 Renewal-Impulses for Culture and Science—Berlin College Course
CW 82 So that the Human Being Can Become a Complete Human Being
CW 83 Western and Eastern World-Contrast. Paths to Understanding It through Anthroposophy
CW 84 What Did the Goetheanum Intend and What Should Anthroposophy Do?

Lectures to the Members of the Anthroposophical Society

CW 88 Concerning the Astral World and Devachan
CW 89 Consciousness—Life—Form. Fundamental Principles of a Spiritual-Scientific Cosmology
CW 90 Participant Notes from the Lectures during the Years 1903-1905
CW 91 Participant Notes from the Lectures during the Years 1903-1905
CW 92 The Occult Truths of Ancient Myths and Sagas
CW 93 The Temple Legend and the Golden Legend
CW 93a Fundamentals of Esotericism
CW 94 Cosmogony. Popular Occultism. The Gospel of John. The Theosophy in the Gospel of John
CW 95 At the Gates of Theosophy
CW 96 Origin-Impulses of Spiritual Science. Christian Esotericism in the Light of New Spirit-Knowledge
CW 97 The Christian Mystery
CW 98 Nature Beings and Spirit Beings—Their Effects in Our Visible World
CW 99 The Theosophy of the Rosicrucians
CW 100 Human Development and Christ-Knowledge
CW 101 Myths and Legends. Occult Signs and Symbols
CW 102 The Working into Human Beings by Spiritual Beings
CW 103 The Gospel of John
CW 104 The Apocalypse of John
CW 104a From the Picture-Script of the Apocalypse of John
CW 105 Universe, Earth, the Human Being: Their Being and Development, as well as Their Reflection in the Connection between Egyptian Mythology and Modern Culture
CW 106 Egyptian Myths and Mysteries in Relation to the Active Spiritual Forces of the Present
CW 107 Spiritual-Scientific Knowledge of the Human Being
CW 108 Answering the Questions of Life and the World through Anthroposophy
CW 109 The Principle of Spiritual Economy in Connection with the Question of Reincarnation. An Aspect of the Spiritual Guidance of Humanity
CW 110 The Spiritual Hierarchies and Their Reflection in the Physical World. Zodiac, Planets and Cosmos

CW 111 Contained in CW 109
CW 112 The Gospel of John in Relation to the Three Other Gospels, Especially the Gospel of Luke
CW 113 The Orient in the Light of the Occident. The Children of Lucifer and the Brothers of Christ
CW 114 The Gospel of Luke
CW 115 Anthroposophy—Psychosophy—Pneumatosophy
CW 116 The Christ-Impulse and the Development of I-Consciousness
CW 117 The Deeper Secrets of the Development of Humanity in Light of the Gospels
CW 118 The Event of the Christ-Appearance in the Etheric World
CW 119 Macrocosm and Microcosm. The Large World and the Small World. Soul-Questions, Life-Questions, Spirit-Questions
CW 120 The Revelation of Karma
CW 121 The Mission of Individual Folk-Souls in Connection with Germanic-Nordic Mythology
CW 122 The Secrets of the Biblical Creation-Story. The Six-Day Work in the First Book of Moses
CW 123 The Gospel of Matthew
CW 124 Excursus in the Area of the Gospel of Mark
CW 125 Paths and Goals of the Spiritual Human Being. Life Questions in the Light of Spiritual Science
CW 126 Occult History. Esoteric Observations of the Karmic Relationships of Personalities and Events of World History
CW 127 The Mission of the New Spiritual Revelation. The Christ-Event as the Middle-Point of Earth Evolution
CW 128 An Occult Physiology
CW 129 Wonders of the World, Trials of the Soul, and Revelations of the Spirit
CW 130 Esoteric Christianity and the Spiritual Guidance of Humanity
CW 131 From Jesus to Christ
CW 132 Evolution from the View Point of the Truth
CW 133 The Earthly and the Cosmic Human Being
CW 134 The World of the Senses and the World of the Spirit
CW 135 Reincarnation and Karma and their Meaning for the Culture of the Present
CW 136 The Spiritual Beings in Celestial Bodies and the Realms of Nature
CW 137 The Human Being in the Light of Occultism, Theosophy and Philosophy
CW 138 On Initiation. On Eternity and the Passing Moment. On the Light of the Spirit and the Darkness of Life
CW 139 The Gospel of Mark
CW 140 Occult Investigation into the Life between Death and New Birth. The Living Interaction between Life and Death
CW 141 Life between Death and New Birth in Relationship to Cosmic Facts

CW 142 The Bhagavad Gita and the Letters of Paul
CW 143 Experiences of the Supersensible. Three Paths of the Soul to Christ
CW 144 The Mysteries of the East and of Christianity
CW 145 What Significance Does Occult Development of the Human Being Have for the Sheaths—Physical Body, Etheric Body, Astral Body, and Self?
CW 146 The Occult Foundations of the Bhagavad Gita
CW 147 The Secrets of the Threshold
CW 148 Out of Research in the Akasha: The Fifth Gospel
CW 149 Christ and the Spiritual World. Concerning the Search for the Holy Grail
CW 150 The World of the Spirit and Its Extension into Physical Existence; The Influence of the Dead in the World of the Living
CW 151 Human Thought and Cosmic Thought
CW 152 Preliminary Stages to the Mystery of Golgotha
CW 153 The Inner Being of the Human Being and Life Between Death and New Birth
CW 154 How Does One Gain an Understanding of the Spiritual World? The Flowing in of Spiritual Impulses from out of the World of the Deceased
CW 155 Christ and the Human Soul. Concerning the Meaning of Life. Theosophical Morality. Anthroposophy and Christianity
CW 156 Occult Reading and Occult Hearing
CW 157 Human Destinies and the Destiny of Peoples
CW 157a The Formation of Destiny and the Life after Death
CW 158 The Connection Between the Human Being and the Elemental World. Kalevala—Olaf Åsteson—The Russian People—The World as the Result of the Influences of Equilibrium
CW 159 The Mystery of Death. The Nature and Significance of Middle Europe and the European Folk Spirits
CW 160 In CW 159
CW 161 Paths of Spiritual Knowledge and the Renewal of the Artistic Worldview
CW 162 Questions of Art and Life in Light of Spiritual Science
CW 163 Coincidence, Necessity and Providence. Imaginative Knowledge and the Processes after Death
CW 164 The Value of Thinking for a Knowledge That Satisfies the Human Being. The Relationship of Spiritual Science to Natural Science
CW 165 The Spiritual Unification of Humanity through the Christ-Impulse
CW 166 Necessity and Freedom in the Events of the World and in Human Action
CW 167 The Present and the Past in the Human Spirit
CW 168 The Connection between the Living and the Dead
CW 169 World-being and Selfhood
CW 170 The Riddle of the Human Being. The Spiritual Background of Human History. Cosmic and Human History, Vol. 1

CW 171 Inner Development-Impulses of Humanity. Goethe and the Crisis of the 19th Century. Cosmic and Human History, Vol. 2

CW 172 The Karma of the Vocation of the Human Being in Connection with Goethe's Life. Cosmic and Human History, Vol. 3

CW 173 Contemporary-Historical Considerations: The Karma of Untruthfulness, Part One. Cosmic and Human History, Vol. 4

CW 174 Contemporary-Historical Considerations: The Karma of Untruthfulness, Part Two. Cosmic and Human History, Vol. 5

CW 174a Middle Europe between East and West. Cosmic and Human History, Vol. 6

CW 174b The Spiritual Background of the First World War. Cosmic and Human History, Vol. 7

CW 175 Building Stones for an Understanding of the Mystery of Golgotha. Cosmic and Human Metamorphoses

CW 176 Truths of Evolution of the Individual and Humanity. The Karma of Materialism

CW 177 The Spiritual Background of the Outer World. The Fall of the Spirits of Darkness. Spiritual Beings and Their Effects, Vol. 1

CW 178 Individual Spiritual Beings and their Influence in the Soul of the Human Being. Spiritual Beings and their Effects, Vol. 2

CW 179 Spiritual Beings and Their Effects. Historical Necessity and Freedom. The Influences on Destiny from out of the World of the Dead. Spiritual Beings and Their Effects, Vol. 3

CW 180 Mystery Truths and Christmas Impulses. Ancient Myths and their Meaning. Spiritual Beings and Their Effects, Vol. 4

CW 181 Earthly Death and Cosmic Life. Anthroposophical Gifts for Life. Necessities of Consciousness for the Present and the Future.

CW 182 Death as Transformation of Life

CW 183 The Science of the Development of the Human Being

CW 184 The Polarity of Duration and Development in Human Life. The Cosmic Pre-History of Humanity

CW 185 Historical Symptomology

CW 185a Historical-Developmental Foundations for Forming a Social Judgement

CW 186 The Fundamental Social Demands of Our Time—In Changed Situations

CW 187 How Can Humanity Find the Christ Again? The Threefold Shadow-Existence of our Time and the New Christ-Light

CW 188 Goetheanism, a Transformation-Impulse and Resurrection-Thought. Science of the Human Being and Science of Sociology

CW 189 The Social Question as a Question of Consciousness. The Spiritual Background of the Social Question, Vol. 1

CW 190 Impulses of the Past and the Future in Social Occurrences. The Spiritual Background of the Social Question, Vol. 2

CW 191	Social Understanding from Spiritual-Scientific Cognition. The Spiritual Background of the Social Question, Vol. 3
CW 192	Spiritual-Scientific Treatment of Social and Pedagogical Questions
CW 193	The Inner Aspect of the Social Riddle. Luciferic Past and Ahrimanic Future
CW 194	The Mission of Michael. The Revelation of the Actual Mysteries of the Human Being
CW 195	Cosmic New Year and the New Year Idea
CW 196	Spiritual and Social Transformations in the Development of Humanity
CW 197	Polarities in the Development of Humanity: West and East Materialism and Mysticism Knowledge and Belief
CW 198	Healing Factors for the Social Organism
CW 199	Spiritual Science as Knowledge of the Foundational Impulses of Social Formation
CW 200	The New Spirituality and the Christ-Experience of the 20th Century
CW 201	The Correspondences Between Microcosm and Macrocosm. The Human Being—A Hieroglyph of the Universe. The Human Being in Relationship with the Cosmos: 1
CW 202	The Bridge between the World-Spirituality and the Physical Aspect of the Human Being. The Search for the New Isis, the Divine Sophia. The Human Being in Relationship with the Cosmos: 2
CW 203	The Responsibility of Human Beings for the Development of the World through their Spiritual Connection with the Planet Earth and the World of the Stars. The Human Being in Relationship with the Cosmos: 3
CW 204	Perspectives of the Development of Humanity. The Materialistic Knowledge-Impulse and the Task of Anthroposophy. The Human Being in Relationship with the Cosmos: 4
CW 205	Human Development, World-Soul, and World-Spirit. Part One: The Human Being as a Being of Body and Soul in Relationship to the World. The Human Being in Relationship with the Cosmos: 5
CW 206	Human Development, World-Soul, and World-Spirit. Part Two: The Human Being as a Spiritual Being in the Process of Historical Development. The Human Being in Relationship with the Cosmos: 6
CW 207	Anthroposophy as Cosmosophy. Part One: Characteristic Features of the Human Being in the Earthly and the Cosmic Realms. The Human Being in Relationship with the Cosmos: 7
CW 208	Anthroposophy as Cosmosophy. Part Two: The Forming of the Human Being as the Result of Cosmic Influence. The Human Being in Relationship with the Cosmos: 8
CW 209	Nordic and Central European Spiritual Impulses. The Festival of the Appearance of Christ. The Human Being in Relationship with the Cosmos: 9

CW 210 Old and New Methods of Initiation. Drama and Poetry in the Change of Consciousness in the Modern Age
CW 211 The Sun Mystery and the Mystery of Death and Resurrection. Exoteric and Esoteric Christianity
CW 212 Human Soul Life and Spiritual Striving in Connection with World and Earth Development
CW 213 Human Questions and World Answers
CW 214 The Mystery of the Trinity: The Human Being in Relationship with the Spiritual World in the Course of Time
CW 215 Philosophy, Cosmology, and Religion in Anthroposophy
CW 216 The Fundamental Impulses of the World-Historical Development of Humanity
CW 217 Spiritually Active Forces in the Coexistence of the Older and Younger Generations. Pedagogical Course for Youth
CW 217a Youth's Cognitive Task
CW 218 Spiritual Connections in the Forming of the Human Organism
CW 219 The Relationship of the World of the Stars to the Human Being, and of the Human Being to the World of the Stars. The Spiritual Communion of Humanity
CW 220 Living Knowledge of Nature. Intellectual Fall and Spiritual Redemption
CW 221 Earth-Knowing and Heaven-Insight
CW 222 The Imparting of Impulses to World-Historical Events through Spiritual Powers
CW 223 The Cycle of the Year as Breathing Process of the Earth and the Four Great Festival-Seasons. Anthroposophy and the Human Heart (*Gemüt*)
CW 224 The Human Soul and its Connection with Divine-Spiritual Individualities. The Internalization of the Festivals of the Year
CW 225 Three Perspectives of Anthroposophy. Cultural Phenomena observed from a Spiritual-Scientific Perspective
CW 226 Human Being, Human Destiny, and World Development
CW 227 Initiation-Knowledge
CW 228 Science of Initiation and Knowledge of the Stars. The Human Being in the Past, the Present, and the Future from the Viewpoint of the Development of Consciousness
CW 229 The Experiencing of the Course of the Year in Four Cosmic Imaginations
CW 230 The Human Being as Harmony of the Creative, Building, and Formative World-Word
CW 231 The Supersensible Human Being, Understood Anthroposophically
CW 232 The Forming of the Mysteries
CW 233 World History Illuminated by Anthroposophy and as the Foundation for Knowledge of the Human Spirit

CW 233a	Mystery Sites of the Middle Ages: Rosicrucianism and the Modern Initiation-Principle. The Festival of Easter as Part of the History of the Mysteries of Humanity
CW 234	Anthroposophy. A Summary after 21 Years
CW 235	Esoteric Observations of Karmic Relationships in 6 Volumes, Vol. 1
CW 236	Esoteric Observations of Karmic Relationships in 6 Volumes, Vol. 2
CW 237	Esoteric Observations of Karmic Relationships in 6 Volumes, Vol. 3: The Karmic Relationships of the Anthroposophical Movement
CW 238	Esoteric Observations of Karmic Relationships in 6 Volumes, Vol. 4: The Spiritual Life of the Present in Relationship to the Anthroposophical Movement
CW 239	Esoteric Observations of Karmic Relationships in 6 Volumes, Vol. 5
CW 240	Esoteric Observations of Karmic Relationships in 6 Volumes, Vol. 6
CW 243	The Consciousness of the Initiate
CW 245	Instructions for an Esoteric Schooling
CW 250	The Building-Up of the Anthroposophical Society. From the Beginning to the Outbreak of the First World War
CW 251	The History of the Goetheanum Building-Association
CW 252	Life in the Anthroposophical Society from the First World War to the Burning of the First Goetheanum
CW 253	The Problems of Living Together in the Anthroposophical Society. On the Dornach Crisis of 1915. With Highlights on Swedenborg's Clairvoyance, the Views of Freudian Psychoanalysts, and the Concept of Love in Relation to Mysticism
CW 254	The Occult Movement in the 19th Century and Its Relationship to World Culture. Significant Points from the Exoteric Cultural Life around the Middle of the 19th Century
CW 255	Rudolf Steiner during the First World War
CW 255a	Anthroposophy and the Reformation of Society. On the History of the Threefold Movement
CW 255b	Anthroposophy and Its Opponents, 1919–1921
CW 256	How Can the Anthroposophical Movement Be Financed?
CW 256a	Futurum, Inc. / International Laboratories, Inc.
CW 256b	The Coming Day, Inc.
CW 257	Anthroposophical Community-Building
CW 258	The History of and Conditions for the Anthroposophical Movement in Relationship to the Anthroposophical Society. A Stimulus to Self-Contemplation
CW 259	The Year of Destiny 1923 in the History of the Anthroposophical Society. From the Burning of the Goetheanum to the Christmas Conference
CW 260	The Christmas Conference for the Founding of the General Anthroposophical Society

CW 260a The Constitution of the General Anthroposophical Society and the School for Spiritual Science. The Rebuilding of the Goetheanum
CW 261 Our Dead. Addresses, Words of Remembrance, and Meditative Verses, 1906-1924
CW 262 Rudolf Steiner and Marie Steiner-von Sivers: Correspondence and Documents, 1901-1925
CW 263/1 Rudolf Steiner and Edith Maryon: Correspondence: Letters, Verses, Sketches, 1912-1924
CW 264 On the History and the Contents of the First Section of the Esoteric School from 1904 to 1914. Letters, Newsletters, Documents, Lectures
CW 265 On the History and from the Contents of the Ritual-Knowledge Section of the Esoteric School from 1904 to 1914. Documents, and Lectures from the Years 1906 to 1914, as well as on New Approaches to Ritual-Knowledge Work in the Years 1921–1924
CW 266/1 From the Contents of the Esoteric Lessons. Volume 1: 1904–1909. Notes from Memory of Participants. Meditation texts from the notes of Rudolf Steiner
CW 266/2 From the Contents of the Esoteric Lessons. Volume 2: 1910–1912. Notes from Memory of Participants
CW 266/3 From the Contents of the Esoteric Lessons. Volume 3: 1913, 1914 and 1920–1923. Notes from Memory of Participants. Meditation texts from the notes of Rudolf Steiner
CW 267 Soul-Exercises: Vol. 1: Exercises with Word and Image Meditations for the Methodological Development of Higher Powers of Knowledge, 1904–1924
CW 268 Soul-Exercises: Vol. 2: Mantric Verses, 1903–1925
CW 269 Ritual Texts for the Celebration of the Free Christian Religious Instruction. The Collected Verses for Teachers and Students of the Waldorf School
CW 270 Esoteric Instructions for the First Class of the School for Spiritual Science at the Goetheanum 1924, 4 Volumes
CW 271 Art and Knowledge of Art. Foundations of a New Aesthetic
CW 272 Spiritual-Scientific Commentary on Goethe's 'Faust' in Two Volumes. Vol. 1: Faust, the Striving Human Being
CW 273 Spiritual-Scientific Commentary on Goethe's 'Faust' in Two Volumes. Vol. 2: The Faust-Problem
CW 274 Addresses for the Christmas Plays from the Old Folk Traditions
CW 275 Art in the Light of Mystery-Wisdom
CW 276 The Artistic in Its Mission in the World. The Genius of Language. The World of Self-Revealing Radiant Appearances—Anthroposophy and Art. Anthroposophy and Poetry
CW 277 Eurythmy. The Revelation of the Speaking Soul
CW 277a The Origin and Development of Eurythmy

CW 278	Eurythmy as Visible Song
CW 279	Eurythmy as Visible Speech
CW 280	The Method and Nature of Speech Formation
CW 281	The Art of Recitation and Declamation
CW 282	Speech Formation and Dramatic Art
CW 283	The Nature of Things Musical and the Experience of Tone in the Human Being
CW 284/285	Images of Occult Seals and Pillars. The Munich Congress of Whitsun 1907 and Its Consequences
CW 286	Paths to a New Style of Architecture. 'And the Building Becomes Human'
CW 287	The Building at Dornach as a Symbol of Historical Becoming and an Artistic Transformation Impulse
CW 288	Style-Forms in the Living Organic
CW 289	The Building-Idea of the Goetheanum: Lectures with Slides from the Years 1920–1921
CW 290	The Building-Idea of the Goetheanum: Lectures with Slides from the Years 1920–1921
CW 291	The Nature of Colours
CW 291a	Knowledge of Colours. Supplementary Volume to 'The Nature of Colours'
CW 292	Art History as Image of Inner Spiritual Impulses
CW 293	General Knowledge of the Human Being as the Foundation of Pedagogy
CW 294	The Art of Education, Methodology and Didactics
CW 295	The Art of Education: Seminar Discussions and Lectures on Lesson Planning
CW 296	The Question of Education as a Social Question
CW 297	The Idea and Practice of the Waldorf School
CW 297a	Education for Life: Self-Education and the Practice of Pedagogy
CW 298	Rudolf Steiner in the Waldorf School
CW 299	Spiritual-Scientific Observations on Speech
CW 300a	Conferences with the Teachers of the Free Waldorf School in Stuttgart, 1919 to 1924, in 3 Volumes, Vol. 1
CW 300b	Conferences with the Teachers of the Free Waldorf School in Stuttgart, 1919 to 1924, in 3 Volumes, Vol. 2
CW 300c	Conferences with the Teachers of the Free Waldorf School in Stuttgart, 1919 to 1924, in 3 Volumes, Vol. 3
CW 301	The Renewal of Pedagogical-Didactical Art through Spiritual Science
CW 302	Knowledge of the Human Being and the Forming of Class Lessons
CW 302a	Education and Teaching from a Knowledge of the Human Being
CW 303	The Healthy Development of the Human Being
CW 304	Methods of Education and Teaching Based on Anthroposophy
CW 304a	Anthroposophical Knowledge of the Human Being and Pedagogy

CW 305 The Soul-Spiritual Foundational Forces of the Art of Education. Spiritual Values in Education and Social Life
CW 306 Pedagogical Praxis from the Viewpoint of a Spiritual-Scientific Knowledge of the Human Being. The Education of the Child and Young Human Beings
CW 307 The Spiritual Life of the Present and Education
CW 308 The Method of Teaching and the Life-Requirements for Teaching
CW 309 Anthroposophical Pedagogy and Its Prerequisites
CW 310 The Pedagogical Value of a Knowledge of the Human Being and the Cultural Value of Pedagogy
CW 311 The Art of Education from an Understanding of the Being of Humanity
CW 312 Spiritual Science and Medicine
CW 313 Spiritual-Scientific Viewpoints on Therapy
CW 314 Physiology and Therapy Based on Spiritual Science
CW 315 Curative Eurythmy
CW 316 Meditative Observations and Instructions for a Deepening of the Art of Healing
CW 317 The Curative Education Course
CW 318 The Working Together of Doctors and Pastors
CW 319 Anthroposophical Knowledge of the Human Being and Medicine
CW 320 Spiritual-Scientific Impulses for the Development of Physics 1: The First Natural-Scientific Course: Light, Colour, Tone, Mass, Electricity, Magnetism
CW 321 Spiritual-Scientific Impulses for the Development of Physics 2: The Second Natural-Scientific Course: Warmth at the Border of Positive and Negative Materiality
CW 322 The Borders of the Knowledge of Nature
CW 323 The Relationship of the various Natural-Scientific Fields to Astronomy
CW 324 Nature Observation, Mathematics, and Scientific Experimentation and Results from the Viewpoint of Anthroposophy
CW 324a The Fourth Dimension in Mathematics and Reality
CW 325 Natural Science and the World-Historical Development of Humanity since Ancient Times
CW 326 The Moment of the Coming Into Being of Natural Science in World History and Its Development Since Then
CW 327 Spiritual-Scientific Foundations for Success in Farming. The Agricultural Course
CW 328 The Social Question
CW 329 The Liberation of the Human Being as the Foundation for a New Social Form
CW 330 The Renewal of the Social Organism
CW 331 Work-Council and Socialization
CW 332 The Alliance for Threefolding and the Total Reform of Society. The Council on Culture and the Liberation of the Spiritual Life

CW 332a	The Social Future
CW 333	Freedom of Thought and Social Forces
CW 334	From the Unified State to the Threefold Social Organism
CW 335	The Crisis of the Present and the Path to Healthy Thinking
CW 336	The Great Questions of the Times and Anthroposophical Spiritual Knowledge
CW 337a	Social Ideas, Social Reality, Social Practice, Vol. 1: Question-and-Answer Evenings and Study Evenings of the Alliance for the Threefold Social Organism in Stuttgart, 1919-1920
CW 337b	Social Ideas, Social Realities, Social Practice, Vol. 2: Discussion Evenings of the Swiss Alliance for the Threefold Social Organism
CW 338	How Does One Work on Behalf of the Impulse for the Threefold Social Organism?
CW 339	Anthroposophy, Threefold Social Organism, and the Art of Public Speaking
CW 340	The National-Economics Course. The Tasks of a New Science of Economics, Volume 1
CW 341	The National-Economics Seminar. The Tasks of a New Science of Economics, Volume 2
CW 342	Lectures and Courses on Christian Religious Work, Vol. 1: Anthroposophical Foundations for a Renewed Christian Religious Working
CW 343	Lectures and Courses on Christian Religious Work, Vol. 2: Spiritual Knowledge—Religious Feeling—Cultic Doing
CW 344	Lectures and Courses on Christian Religious Work, Vol. 3: Lectures at the Founding of the Christian Community
CW 345	Lectures and Courses on Christian Religious Work, Vol. 4: Concerning the Nature of the Working Word
CW 346	Lectures and Courses on Christian Religious Work, Vol. 5: The Apocalypse and the Work of the Priest
CW 347	The Knowledge of the Nature of the Human Being According to Body, Soul and Spirit. On Earlier Conditions of the Earth
CW 348	On Health and Illness. Foundations of a Spiritual-Scientific Doctrine of the Senses
CW 349	On the Life of the Human Being and of the Earth. On the Nature of Christianity
CW 350	Rhythms in the Cosmos and in the Human Being. How Does One Come To See the Spiritual World?
CW 351	The Human Being and the World. The Influence of the Spirit in Nature. On the Nature of Bees
CW 352	Nature and the Human Being Observed Spiritual-Scientifically
CW 353	The History of Humanity and the World-Views of the Folk Cultures
CW 354	The Creation of the World and the Human Being. Life on Earth and the Influence of the Stars

SIGNIFICANT EVENTS IN THE LIFE OF RUDOLF STEINER

1829: June 23: birth of Johann Steiner (1829–1910)—Rudolf Steiner's father—in Geras, Lower Austria.

1834: May 8: birth of Franciska Blie (1834–1918)—Rudolf Steiner's mother—in Horn, Lower Austria. 'My father and mother were both children of the glorious Lower Austrian forest district north of the Danube.'

1860: May 16: marriage of Johann Steiner and Franciska Blie.

1861: February 25: birth of *Rudolf Joseph Lorenz Steiner* in Kraljevec, Croatia, near the border with Hungary, where Johann Steiner works as a telegrapher for the South Austria Railroad. Rudolf Steiner is baptized two days later, February 27, the date usually given as his birthday.

1862: Summer: the family moves to Modling, Lower Austria.

1863: The family moves to Pottschach, Lower Austria, near the Styrian border, where Johann Steiner becomes station master. 'The view stretched to the mountains . . . majestic peaks in the distance and the sweet charm of nature in the immediate surroundings.'

1864: November 15: birth of Rudolf Steiner's sister, Leopoldine (d. November 1, 1927). She will become a seamstress and live with her parents for the rest of her life.

1866: July 28: birth of Rudolf Steiner's deaf-mute brother, Gustav (d. May 1, 1941).

1867: Rudolf Steiner enters the village school. Following a disagreement between his father and the schoolmaster, whose wife falsely accused the boy of causing a commotion, Rudolf Steiner is taken out of school and taught at home.

1868: A critical experience. Unknown to the family, an aunt dies in a distant town. Sitting in the station waiting room, Rudolf Steiner sees her 'form,' which speaks to him, asking for help. 'Beginning with this experience, a new soul life began in the boy, one in which not only the outer trees and mountains spoke to him, but also the worlds that lay behind them. From this moment on, the boy began to live with the spirits of nature . . .'

1869: The family moves to the peaceful, rural village of Neudorfl, near Wiener Neustadt in present-day Austria. Rudolf Steiner attends the village school. Because of the 'unorthodoxy' of his writing and spelling, he has to do 'extra lessons'.

1870: Through a book lent to him by his tutor, he discovers geometry: 'To grasp something purely in the spirit brought me inner happiness. I know that I first learned happiness through geometry.' The same tutor allows him to draw, while other students still struggle with their reading and writing. 'An artistic element' thus enters his education.

1871: Though his parents are not religious, Rudolf Steiner becomes a 'church child,' a favourite of the priest, who was 'an exceptional character.' 'Up to the age of ten or eleven, among those I came to know, he was far and away the most significant.' Among other things, he introduces Steiner to Copernican, heliocentric cosmology. As an altar boy, Rudolf Steiner serves at masses, funerals, and Corpus Christi processions. At year's end, after an incident in which he escapes a thrashing, his father forbids him to go to church.

1872: Rudolf Steiner transfers to grammar school in Wiener-Neustadt, a five-mile walk from home, which must be done in all weathers.

1873–75: Through his teachers and on his own, Rudolf Steiner has many wonderful experiences with science and mathematics. Outside school, he teaches himself analytic geometry, trigonometry, differential equations, and calculus.

1876: Rudolf Steiner begins tutoring other students. He learns bookbinding from his father. He also teaches himself stenography.

1877: Rudolf Steiner discovers Kant's *Critique of Pure Reason,* which he reads and rereads. He also discovers and reads von Rotteck's *World History.*

1878: He studies extensively in contemporary psychology and philosophy.

1879: Rudolf Steiner graduates from high school with honours. His father is transferred to Inzersdorf, near Vienna. He uses his first visit to Vienna 'to purchase a great number of philosophy books'—Kant, Fichte, Schelling, and Hegel, as well as numerous histories of philosophy. His aim: to find a path from the 'I' to nature.

October 1879–1883: Rudolf Steiner attends the Technical College in Vienna—to study mathematics, chemistry, physics, mineralogy, botany, zoology, biology, geology, and mechanics—with a scholarship. He also attends lectures in history and literature, while avidly reading philosophy on his own. His two favourite professors are Karl Julius Schröer (German language and literature) and Edmund Reitlinger (physics). He also audits lectures by Robert Zimmermann on aesthetics and Franz Brentano on philosophy. During this year he begins his friendship with Moritz Zitter (1861–1921), who will help support him financially when he is in Berlin.

1880: Rudolf Steiner attends lectures on Schiller and Goethe by Karl Julius Schröer, who becomes his mentor. Also 'through a remarkable combination of circumstances,' he meets Felix Koguzki, a 'herb gatherer' and healer, who could 'see deeply into the secrets of nature'. Rudolf Steiner will meet and study with this 'emissary of the Master' throughout his time in Vienna.

1881: January: '... I didn't sleep a wink. I was busy with philosophical problems until about 12:30 a.m. Then, finally, I threw myself down on my couch. All my striving during the previous year had been to research whether the following statement by Schelling was true or not: *Within everyone dwells a secret, marvellous capacity to draw back from the stream of time—out of the self clothed in all that comes to us from outside—into our innermost being and there, in the immutable form of the Eternal, to look into ourselves.* I believe, and I am still quite certain of it, that I discovered this capacity in myself; I had long had an inkling of it. Now the whole of idealist philosophy stood before me in modified form. What's a sleepless night compared to that!'

Rudolf Steiner begins communicating with leading thinkers of the day, who send him books in return, which he reads eagerly.

July: 'I am not one of those who dives into the day like an animal in human form. I pursue a quite specific goal, an idealistic aim—knowledge of the truth! This cannot be done offhandedly. It requires the greatest striving in the world, free of all egotism, and equally of all resignation.'

August: Steiner puts down on paper for the first time thoughts for a 'Philosophy of Freedom.' 'The striving for the absolute: this human yearning is freedom.' He also seeks to outline a 'peasant philosophy,' describing what the worldview of a 'peasant'—one who lives close to the earth and the old ways—really is.

1881–1882: Felix Koguzki, the herb gatherer, reveals himself to be the envoy of another, higher initiatory personality, who instructs Rudolf Steiner to penetrate Fichte's philosophy and to master modern scientific thinking as a preparation for right entry into the spirit. This 'Master' also teaches him the double (evolutionary and involutionary) nature of time.

1882: Through the offices of Karl Julius Schröer, Rudolf Steiner is asked by Joseph Kürschner to edit Goethe's scientific works for the *Deutschen National-Literatur* edition. He writes 'A Possible Critique of Atomistic Concepts' and sends it to Friedrich Theodor Vischer.

1883: Rudolf Steiner completes his college studies and begins work on the Goethe project.

1884: First volume of Goethe's *Scientific Writings* (CW 1) appears (March). He lectures on Goethe and Lessing, and Goethe's approach to science. In July, he enters the household of Ladislaus and Pauline Specht as tutor to the four Specht boys. He will live there until 1890. At this time, he meets Josef Breuer (1842–1925), the co-author with Sigmund Freud of *Studies in Hysteria,* who is the Specht family doctor.

1885: While continuing to edit Goethe's writings, Rudolf Steiner reads deeply in contemporary philosophy (Eduard von Hartmann, Johannes Volkelt, and Richard Wahle, among others).

1886: May: Rudolf Steiner sends Kürschner the manuscript of *Outlines of Goethe's Theory of Knowledge* (CW 2), which appears in October, and which he sends out widely. He also meets the poet Marie Eugenie Delle Grazie and writes 'Nature and Our Ideals' for her. He attends her salon, where he meets many priests, theologians, and philosophers, who will become his friends. Meanwhile, the director of the Goethe Archive in Weimar requests his collaboration with the *Sophien* edition of Goethe's works, particularly the writings on colour.

1887: At the beginning of the year, Rudolf Steiner is very sick. As the year progresses and his health improves, he becomes increasingly 'a man of letters,' lecturing, writing essays, and taking part in Austrian cultural life. In August–September, the second volume of Goethe's *Scientific Writings* appears.

1888: January–July: Rudolf Steiner assumes editorship of the 'German Weekly' *(Deutsche Wochenschrift)*. He begins lecturing more intensively, giving, for example, a lecture titled 'Goethe as Father of a New Aesthetics.' He meets and becomes soul friends with Friedrich Eckstein (1861–1939), a vegetarian, philosopher of symbolism, alchemist, and musician, who will introduce him to various spiritual currents (including Theosophy) and with whom he will meditate and interpret esoteric and alchemical texts.

1889: Rudolf Steiner first reads Nietzsche *(Beyond Good and Evil)*. He encounters Theosophy again and learns of Madame Blavatsky in the theosophical circle around Marie Lang (1858–1934). Here he also meets well-known figures of Austrian life, as well as esoteric figures like the occultist Franz Hartmann and Karl Leinigen-Billigen (translator of C.G. Harrison's *The Transcendental Universe)*. During this period, Steiner first reads A.P. Sinnett's *Esoteric Buddhism* and Mabel Collins's *Light on the Path*. He also begins travelling, visiting Budapest, Weimar, and Berlin (where he meets philosopher Eduard von Hartmann).

1890: Rudolf Steiner finishes Volume 3 of Goethe's scientific writings. He begins his doctoral dissertation, which will become *Truth and Science* (CW 3). He also meets the poet and feminist Rosa Mayreder (1858–1938), with whom he can exchange his most intimate thoughts. In September, Rudolf Steiner moves to Weimar to work in the Goethe-Schiller Archive.

1891: Volume 3 of the Kürschner edition of Goethe appears. Meanwhile, Rudolf Steiner edits Goethe's studies in mineralogy and scientific writings for the *Sophien* edition. He meets Ludwig Laistner of the Cotta Publishing Company, who asks for a book on the basic question of metaphysics. From this will result, ultimately, *The Philosophy of Freedom* (CW 4), which will be published not by Cotta but by Emil Felber. In October, Rudolf Steiner takes the oral exam for a doctorate in philosophy, mathematics, and mechanics at Rostock University, receiving his doctorate on the twenty-sixth. In November, he gives his first lecture on Goethe's 'Fairy Tale' in Vienna.

1892: Rudolf Steiner continues work at the Goethe-Schiller Archive and on his *Philosophy of Freedom*. *Truth and Science*, his doctoral dissertation, is published. Steiner undertakes to write Introductions to books on Schopenhauer and Jean Paul for Cotta. At year's end, he finds lodging with Anna Eunike, née Schulz (1853–1911), a widow with four daughters and a son. He also develops a friendship with Otto Erich Hartleben (1864–1905) with whom he shares literary interests.

1893: Rudolf Steiner begins his habit of producing many reviews and articles. In March, he gives a lecture titled 'Hypnotism, with Reference to Spiritism.' In September, volume 4 of the Kürschner edition is completed. In November, *The Philosophy of Freedom* appears. This year, too, he meets John Henry Mackay (1864–1933), the anarchist, and Max Stirner, a scholar and biographer.

1894: Rudolf Steiner meets Elisabeth Fürster Nietzsche, the philosopher's sister, and begins to read Nietzsche in earnest, beginning with the as yet unpublished *Antichrist*. He also meets Ernst Haeckel (1834–1919). In the fall, he begins to write *Nietzsche, A Fighter against His Time* (CW 5).

1895: May, *Nietzsche, A Fighter against His Time* appears.

1896: January 22: Rudolf Steiner sees Friedrich Nietzsche for the first and only time. Moves between the Nietzsche and the Goethe-Schiller

Archives, where he completes his work before year's end. He falls out with Elisabeth Förster Nietzsche, thus ending his association with the Nietzsche Archive.

1897: Rudolf Steiner finishes the manuscript of *Goethe's Worldview* (CW 6). He moves to Berlin with Anna Eunike and begins editorship of the *Magazin für Literatur*. From now on, Steiner will write countless reviews, literary and philosophical articles, and so on. He begins lecturing at the 'Free Literary Society.' In September, he attends the Zionist Congress in Basel. He sides with Dreyfus in the Dreyfus affair.

1898: Rudolf Steiner is very active as an editor in the political, artistic, and theatrical life of Berlin. He becomes friendly with John Henry Mackay and poet Ludwig Jacobowski (1868–1900). He joins Jacobowski's circle of writers, artists, and scientists—'The Coming Ones' (*Die Kommenden*)—and contributes lectures to the group until 1903. He also lectures at the 'League for College Pedagogy.' He writes an article for Goethe's sesquicentennial, 'Goethe's Secret Revelation,' on the 'Fairy Tale of the Green Snake and the Beautiful Lily.'

1898–99: 'This was a trying time for my soul as I looked at Christianity. . . . I was able to progress only by contemplating, by means of spiritual perception, the evolution of Christianity. . . . Conscious knowledge of real Christianity began to dawn in me around the turn of the century. This seed continued to develop. My soul trial occurred shortly before the beginning of the twentieth century. It was decisive for my soul's development that I stood spiritually before the Mystery of Golgotha in a deep and solemn celebration of knowledge.'

1899: Rudolf Steiner begins teaching and giving lectures and lecture cycles at the Workers' College, founded by Wilhelm Liebknecht (1826–1900). He will continue to do so until 1904. Writes: *Literature and Spiritual Life in the Nineteenth Century; Individualism in Philosophy; Haeckel and His Opponents; Poetry in the Present;* and begins what will become (fifteen years later) *The Riddles of Philosophy* (CW 18). He also meets many artists and writers, including Käthe Kollwitz, Stefan Zweig, and Rainer Maria Rilke. On October 31, he marries Anna Eunike.

1900: 'I thought that the turn of the century must bring humanity a new light. It seemed to me that the separation of human thinking and willing from the spirit had peaked. A turn or reversal of direction in human evolution seemed to me a necessity.' Rudolf Steiner finishes *World and Life Views in the Nineteenth Century* (the second part of what will become *The Riddles of Philosophy)* and dedicates it to

Ernst Haeckel. It is published in March. He continues lecturing at *Die Kommenden,* whose leadership he assumes after the death of Jacobowski. Also, he gives the Gutenberg Jubilee lecture before 7,000 typesetters and printers. In September, Rudolf Steiner is invited by Count and Countess Brockdorff to lecture in the Theosophical Library. His first lecture is on Nietzsche. His second lecture is titled 'Goethe's Secret Revelation.' October 6, he begins a lecture cycle on the mystics that will become *Mystics after Modernism* (CW 7). November–December: 'Marie von Sivers appears in the audience. . . .' Also in November, Steiner gives his first lecture at the Giordano Bruno Bund (where he will continue to lecture until May, 1905). He speaks on Bruno and modern Rome, focusing on the importance of the philosophy of Thomas Aquinas as monism.

1901: In continual financial straits, Rudolf Steiner's early friends Moritz Zitter and Rosa Mayreder help support him. In October, he begins the lecture cycle *Christianity as Mystical Fact* (CW 8) at the Theosophical Library. In November, he gives his first 'theosophical lecture' on Goethe's 'Fairy Tale' in Hamburg at the invitation of Wilhelm Hubbe-Schleiden. He also attends a gathering to celebrate the founding of the Theosophical Society at Count and Countess Brockdorff's. He gives a lecture cycle, 'From Buddha to Christ,' for the circle of the *Kommenden.* November 17, Marie von Sivers asks Rudolf Steiner if Theosophy needs a Western–Christian spiritual movement (to complement Theosophy's Eastern emphasis). 'The question was posed. Now, following spiritual laws, I could begin to give an answer. . . .' In December, Rudolf Steiner writes his first article for a theosophical publication. At year's end, the Brockdorffs and possibly Wilhelm Hubbe-Schleiden ask Rudolf Steiner to join the Theosophical Society and undertake the leadership of the German section. Rudolf Steiner agrees, on the condition that Marie von Sivers (then in Italy) work with him.

1902: Beginning in January, Rudolf Steiner attends the opening of the Workers' School in Spandau with Rosa Luxemberg (1870–1919). January 17, Rudolf Steiner joins the Theosophical Society. In April, he is asked to become general secretary of the German Section of the theosophical Society, and works on preparations for its founding. In July, he visits London for a theosophical congress. He meets Bertram Keightly, G.R.S. Mead, A.P. Sinnett, and Annie Besant, among others. In September, *Christianity as Mystical Fact* appears. In October, Rudolf Steiner gives his first public lecture on Theosophy ('Monism and Theosophy') to about three hundred people at the Giordano Bruno Bund. On October 19–21, the

German Section of the Theosophical Society has its first meeting; Rudolf Steiner is the general secretary, and Annie Besant attends. Steiner lectures on practical karma studies. On October 23, Annie Besant inducts Rudolf Steiner into the Esoteric School of the Theosophical Society. On October 25, Steiner begins a weekly series of lectures: 'The Field of Theosophy.' During this year, Rudolf Steiner also first meets Ita Wegman (1876–1943), who will become his close collaborator in his final years.

1903: Rudolf Steiner holds about 300 lectures and seminars. In May, the first issue of the periodical *Luzifer* appears. In June, Rudolf Steiner visits London for the first meeting of the Federation of the European Sections of the Theosophical Society, where he meets Colonel Olcott. He begins to write *Theosophy* (CW 9).

1904: Rudolf Steiner continues lecturing at the Workers' College and elsewhere (about 90 lectures), while lecturing intensively all over Germany among theosophists (about 140 lectures). In February, he meets Carl Unger (1878–1929), who will become a member of the board of the Anthroposophical Society (1913). In March, he meets Michael Bauer (1871–1929), a Christian mystic, who will also be on the board. In May, *Theosophy* appears, with the dedication: 'To the spirit of Giordano Bruno.' Rudolf Steiner and Marie von Sivers visit London for meetings with Annie Besant. June: Rudolf Steiner and Marie von Sivers attend the meeting of the Federation of European Sections of the Theosophical Society in Amsterdam. In July, Steiner begins the articles in *Luzifer-Gnosis* that will become *How to Know Higher Worlds* (CW 10) and *Cosmic Memory* (CW 11). In September, Annie Besant visits Germany. In December, Steiner lectures on Freemasonry. He mentions the High Grade Masonry derived from John Yarker and represented by Theodore Reuss and Karl Kellner as a blank slate 'into which a good image could be placed'.

1905: This year, Steiner ends his non-theosophical lecturing activity. Supported by Marie von Sivers, his theosophical lecturing—both in public and in the Theosophical Society—increases significantly: 'The German Theosophical Movement is of exceptional importance.' Steiner recommends reading, among others, Fichte, Jacob Boehme, and Angelus Silesius. He begins to introduce Christian themes into Theosophy. He also begins to work with doctors (Felix Peipers and Ludwig Noll). In July, he is in London for the Federation of European Sections, where he attends a lecture by Annie Besant: 'I have seldom seen Mrs Besant speak in so inward and heartfelt a manner... Through Mrs Besant I have found the way to H.P. Blavatsky.' September to October, he gives a course of 31 lectures for a small group of esoteric students. In October, the annual meeting of the German Section of the Theosophical Society, which still remains very small, takes place. Rudolf Steiner reports membership has risen from 121 to 377 members. In November, seeking to establish esoteric 'continuity,' Rudolf Steiner and Marie von Sivers participate in a 'Memphis-Misraim' Masonic ceremony. They pay 45 marks for membership. 'Yesterday, you saw how little remains of former esoteric institutions.' 'We are dealing only with a "framework" . . for the present, nothing lies behind it. The occult powers have completely withdrawn.'

1906: Expansion of theosophical work. Rudolf Steiner gives about 245 lectures, only 44 of which take place in Berlin. Cycles are given in Paris, Leipzig, Stuttgart, and Munich. Esoteric work also intensifies. Rudolf Steiner begins writing *An Outline of Esoteric Science* (CW 13). In January, Rudolf Steiner receives permission (a patent) from the Great Orient of the Scottish A & A Thirty-Three Degree Rite of the Order of the Ancient Freemasons of the Memphis-Misraim Rite to direct a chapter under the name 'Mystica Aeterna.' This will become the 'Cognitive-Ritual Section' (also called 'Misraim Service') of the Esoteric School. (See: *Freemasonry and Ritual Work: The Misraim Service,* CW 265.) During this time, Steiner also meets Albert Schweitzer. In May, he is in Paris, where he visits Édouard Schuré. Many Russians attend his lectures (including Konstantin Balmont, Dimitri Mereszkovski, Zinaida Hippius, and Maximilian Woloshin). He attends the General Meeting of the European Federation of the Theosophical Society, at which Col. Olcott is present for the last time. He spends the year's end in Venice and Rome, where he writes and works on his translation of H.P. Blavatsky's *Key to Theosophy.*

1907: Further expansion of the German Theosophical Movement according to the Rosicrucian directive to 'introduce spirit into the world'—in education, in social questions, in art, and in science. In February, Col. Olcott dies in Adyar. Before he dies, Olcott indicates that 'the Masters' wish Annie Besant to succeed him: much politicking ensues. Rudolf Steiner supports Besant's candidacy. April–May: preparations for the Congress of the Federation of European Sections of the Theosophical Society—the great, watershed Whitsun 'Munich Congress,' attended by Annie Besant and others. Steiner decides to separate Eastern and Western (Christian–Rosicrucian) esoteric schools. He takes his esoteric school out of the Theosophical Society (Besant and Rudolf Steiner are 'in harmony' on this). Steiner makes his first lecture tours to Austria and Hungary. That summer, he is in Italy. In September, he visits Édouard Schuré, who will write the Introduction to the French edition of *Christianity as Mystical Fact* in Barr, Alsace. Rudolf Steiner writes the autobiographical statement known as the 'Barr Document.' In *Luzifer-Gnosis*, 'The Education of the Child' appears.

1908: The movement grows (membership: 1,150). Lecturing expands. Steiner makes his first extended lecture tour to Holland and Scandinavia, as well as visits to Naples and Sicily. Themes: St John's Gospel, the Apocalypse, Egypt, science, philosophy, and logic. *Luzifer-Gnosis* ceases publication. In Berlin, Marie von Sivers (with Johanna Mücke (1864–1949) forms the *Philosophisch-Theosophisch* (after 1915 *Philosophisch-Anthroposophisch) Verlag* to publish Steiner's work. Steiner gives lecture cycles titled *The Gospel of St John* (CW 103) and *The Apocalypse* (104).

1909: *An Outline of Esoteric Science* appears. Lecturing and travel continues. Rudolf Steiner's spiritual research expands to include the polarity of Lucifer and Ahriman; the work of great individualities in history; the Maitreya Buddha and the Bodhisattvas; spiritual economy (CW 109); the work of the spiritual hierarchies in heaven and on earth (CW 110). He also deepens and intensifies his research into the Gospels, giving lectures on the Gospel of St Luke (CW 114) with the first mention of two Jesus children. Meets and becomes friends with Christian Morgenstern (1871–1914). In April, he lays the foundation stone for the Malsch model—the building that will lead to the first Goetheanum. In May, the International Congress of the Federation of European Sections of the Theosophical Society takes place in Budapest. Rudolf Steiner receives the Subba Row medal for *How to Know Higher Worlds.* During this time, Charles W. Leadbeater discovers Jiddu Krishnamurti (1895–1986) and proclaims him the future 'world teacher,' the bearer of the Maitreya Buddha and the 'reappearing Christ.' In October, Steiner delivers seminal lectures on 'anthroposophy,' which he will try, unsuccessfully, to rework over the next years into the unfinished work, *Anthroposophy (A Fragment)* (CW 45).

1910: New themes: *The Reappearance of Christ in the Etheric* (CW 118); *The Fifth Gospel; The Mission of Folk Souls* (CW 121); *Occult History* (CW 126); the evolving development of etheric cognitive capacities. Rudolf Steiner continues his Gospel research with *The Gospel of St Matthew* (CW 123). In January, his father dies. In April, he takes a month-long trip to Italy, including Rome, Monte Cassino, and Sicily. He also visits Scandinavia again. July–August, he writes the first Mystery Drama, *The Portal of Initiation* (CW 14). In November, he gives 'psychosophy' lectures. In December, he submits 'On the

Psychological Foundations and Epistemological Framework of Theosophy' to the International Philosophical Congress in Bologna.

1911: The crisis in the Theosophical Society deepens. In January, 'The Order of the Rising Sun,' which will soon become 'The Order of the Star in the East,' is founded for the coming world teacher, Krishnamurti. At the same time, Marie von Sivers, Rudolf Steiner's co-worker, falls ill. Fewer lectures are given, but important new ground is broken. In Prague, in March, Steiner meets Franz Kafka (1883–1924) and Hugo Bergmann (1883–1975). In April, he delivers his paper to the Philosophical Congress. He writes the second Mystery Drama, *The Soul's Probation* (CW 14). Also, while Marie von Sivers is convalescing, Rudolf Steiner begins work on *Calendar 1912/1913*, which will contain the 'Calendar of the Soul' meditations. On March 19, Anna (Eunike) Steiner dies. In September, Rudolf Steiner visits Einsiedeln, birthplace of Paracelsus. In December, Friedrich Rittelmeyer, future founder of the Christian Community, meets Rudolf Steiner. The *Johannes-Bauverein,* the 'building committee,' which would lead to the first Goetheanum (first planned for Munich), is also founded, and a preliminary committee for the founding of an independent association is created that, in the following year, will become the Anthroposophical Society. Important lecture cycles include *Occult Physiology* (CW 128); *Wonders of the World* (CW 129); *From Jesus to Christ* (CW 131). Other themes: esoteric Christianity; Christian Rosenkreutz; the spiritual guidance of humanity; the sense world and the world of the spirit.

1912: Despite the ongoing, now increasing crisis in the Theosophical Society, much is accomplished: *Calendar 1912/1913* is published; eurythmy is created; both the third Mystery Drama, *The Guardian of the Threshold* (CW 14) and *A Way of Self-Knowledge* (CW 16) are written. New (or renewed) themes included life between death and rebirth and karma and reincarnation. Other lecture cycles: *Spiritual Beings in the Heavenly Bodies and in the Kingdoms of Nature* (CW 136); *The Human Being in the Light of Occultism, Theosophy, and Philosophy* (CW 137); *The Gospel of St Mark* (CW 139); and *The Bhagavad Gita and the Epistles of Paul* (CW 142). On May 8, Rudolf Steiner celebrates White Lotus Day, H.P. Blavatsky's death day, which he had faithfully observed for the past decade, for the last time. In August, Rudolf Steiner suggests the 'independent association' be called the 'Anthroposophical Society.' In September, the first eurythmy course takes place. In October, Rudolf Steiner declines recognition of a Theosophical Society lodge dedicated to the Star of the East and decides to expel all Theosophical Society members belonging to the order.

Also, with Marie von Sivers, he first visits Dornach, near Basel, Switzerland, and they stand on the hill where the Goetheanum will be built. In November, a Theosophical Society lodge is opened by direct mandate from Adyar (Annie Besant). In December, a meeting of the German section occurs at which it is decided that belonging to the Order of the Star of the East is incompatible with membership in the Theosophical Society. December 28: informal founding of the Anthroposophical Society in Berlin.

1913: Expulsion of the German section from the Theosophical Society. February 2–3: Foundation meeting of the Anthroposophical Society. Board members include: Marie von Sivers, Michael Bauer, and Carl Unger. September 20: Laying of the foundation stone for the *Johannes Bau* (Goetheanum) in Dornach. Building begins immediately. The fourth Mystery Drama, *The Soul's Awakening* (CW 14), is completed. Also: *The Threshold of the Spiritual World* (CW 147). Lecture cycles include: *The Bhagavad Gita and the Epistles of Paul* and *The Esoteric Meaning of the Bhagavad Gita* (CW 146), which the Russian philosopher Nikolai Berdyaev attends; *The Mysteries of the East and of Christianity* (CW 144); *The Effects of Esoteric Development* (CW 145); and *The Fifth Gospel* (CW 148). In May, Rudolf Steiner is in London and Paris, where anthroposophical work continues.

1914: Building continues on the *Johannes Bau* (Goetheanum) in Dornach, with artists and co-workers from seventeen nations. The general assembly of the Anthroposophical Society takes place. In May, Rudolf Steiner visits Paris, as well as Chartres Cathedral. June 28: assassination in Sarajevo ('Now the catastrophe has happened!'). August 1: War is declared. Rudolf Steiner returns to Germany from Dornach—he will travel back and forth. He writes the last chapter of *The Riddles of Philosophy*. Lecture cycles include: *Human and Cosmic Thought* (CW 151); *Inner Being of Humanity between Death and a New Birth* (CW 153); *Occult Reading and Occult Hearing* (CW 156). December 24: marriage of Rudolf Steiner and Marie von Sivers.

1915: Building continues. Life after death becomes a major theme, also art. Writes: *Thoughts during a Time of War* (CW 24). Lectures include: *The Secret of Death* (CW 159); *The Uniting of Humanity through the Christ Impulse* (CW 165).

1916: Rudolf Steiner begins work with Edith Maryon (1872–1924) on the sculpture 'The Representative of Humanity' ('The Group'—Christ, Lucifer, and Ahriman). He also works with the alchemist Alexander von Bernus on the quarterly *Das Reich*. He writes *The Riddle of Humanity* (CW 20). Lectures include: *Necessity and Freedom in World History and Human Action* (CW 166); *Past and Present in the*

Human Spirit (CW 167); *The Karma of Vocation* (CW 172); *The Karma of Untruthfulness* (CW 173).

1917: Russian Revolution. The U.S. enters the war. Building continues. Rudolf Steiner delineates the idea of the 'threefold nature of the human being' (in a public lecture March 15) and the 'threefold nature of the social organism' (hammered out in May–June with the help of Otto von Lerchenfeld and Ludwig Polzer-Hoditz in the form of two documents titled *Memoranda,* which were distributed in high places). August–September: Rudolf Steiner writes *The Riddles of the Soul* (CW 20). Also: commentary on 'The Chymical Wedding of Christian Rosenkreutz' for Alexander Bernus (Das *Reich*). Lectures include: *The Karma of Materialism* (CW 176); *The Spiritual Background of the Outer World: The Fall of the Spirits of Darkness* (CW 177).

1918: March 18: peace treaty of Brest-Litovsk—'Now everything will truly enter chaos! What is needed is cultural renewal.' June: Rudolf Steiner visits Karlstein (Grail) Castle outside Prague. Lecture cycle: *From Symptom to Reality in Modern History* (CW 185). In mid-November, Emil Molt, of the Waldorf-Astoria Cigarette Company, has the idea of founding a school for his workers' children.

1919: Focus on the threefold social organism: tireless travel, countless lectures, meetings, and publications. At the same time, a new public stage of Anthroposophy emerges as cultural renewal begins. The coming years will see initiatives in pedagogy, medicine, pharmacology, and agriculture. January 27: threefold meeting: 'We must first of all, with the money we have, found free schools that can bring people what they need.' February: first public eurythmy performance in Zurich. Also: 'Appeal to the German People' (CW 24), circulated March 6 as a newspaper insert. In April, *Towards Social Renewal* (CW 23) appears—'perhaps the most widely read of all books on politics appearing since the war'. Rudolf Steiner is asked to undertake the 'direction and leadership' of the school founded by the Waldorf-Astoria Company. Rudolf Steiner begins to talk about the 'renewal' of education. May 30: a building is selected and purchased for the future Waldorf School. August–September, Rudolf Steiner gives a lecture course for Waldorf teachers, *The Foundations of Human Experience (Study of Man)* (CW 293). September 7: Opening of the first Waldorf School. December (into January): first science course, the *Light Course* (CW 320).

1920: The Waldorf School flourishes. New threefold initiatives. Founding of limited companies *Der Kommende Tag* and *Futurum A.G.* to infuse spiritual values into the economic realm. Rudolf Steiner also focuses on the sciences. Lectures: *Introducing Anthroposophical*

Medicine (CW 312); *The Warmth Course* (CW 321); *The Boundaries of Natural Science* (CW 322); *The Redemption of Thinking* (CW 74). February: Johannes Werner Klein—later a co-founder of the Christian Community—asks Rudolf Steiner about the possibility of a 'religious renewal,' a 'Johannine church.' In March, Rudolf Steiner gives the first course for doctors and medical students. In April, a divinity student asks Rudolf Steiner a second time about the possibility of religious renewal. September 27–October 16: anthroposophical 'university course.' December: lectures titled *The Search for the New Isis* (CW 202).

1921: Rudolf Steiner continues his intensive work on cultural renewal, including the uphill battle for the threefold social order. 'University' arts, scientific, theological, and medical courses include: *The Astronomy Course* (CW 323); *Observation, Mathematics, and Scientific Experiment* (CW 324); the *Second Medical Course* (CW 313); *Colour*. In June and September–October, Rudolf Steiner also gives the first two 'priests' courses' (CW 342 and 343). The 'youth movement' gains momentum. Magazines are founded: *Die Drei* (January), and—under the editorship of Albert Steffen (1884–1963)—the weekly, *Das Goetheanum* (August). In February–March, Rudolf Steiner takes his first trip outside Germany since the war (Holland). On April 7, Steiner receives a letter regarding 'religious renewal,' and May 22–23, he agrees to address the question in a practical way. In June, the Klinical-Therapeutic Institute opens in Arlesheim under the direction of Dr Ita Wegman. In August, the Chemical-Pharmaceutical Laboratory opens in Arlesheim (Oskar Schmiedel and Ita Wegman are directors). The Clinical Therapeutic Institute is inaugurated in Stuttgart (Dr Ludwig Noll is director); also the Research Laboratory in Dornach (Ehrenfried Pfeiffer and Gunther Wachsmuth are directors). In November–December, Rudolf Steiner visits Norway.

1922: The first half of the year involves very active public lecturing (thousands attend); in the second half, Rudolf Steiner begins to withdraw and turn toward the Society—'The Society is asleep.' It is 'too weak' to do what is asked of it. The businesses—*Der Kommende Tag* and *Futurum A.G.*—fail. In January, with the help of an agent, Steiner undertakes a twelve-city German lecture tour, accompanied by eurythmy performances. In two weeks he speaks to more than 2,000 people. In April, he gives a 'university course' in The Hague. He also visits England. In June, he is in Vienna for the East–West Congress. In August–September, he is back in England for the Oxford Conference on Education. Returning to Dornach, he gives the lectures *Philosophy, Cosmology, and Religion*

(CW 215), and gives the third priests' course (CW 344). On September 16, The Christian Community is founded. In October–November, Steiner is in Holland and England. He also speaks to the youth: *The Youth Course* (CW 217). In December, Steiner gives lectures titled *The Origins of Natural Science* (CW 326), and *Humanity and the World of Stars: The Spiritual Communion of Humanity* (CW 219). December 31: Fire at the Goetheanum, which is destroyed.

1923: Despite the fire, Rudolf Steiner continues his work unabated. A very hard year. Internal dispersion, dissension, and apathy abound. There is conflict—between old and new visions—within the Society. A wake-up call is needed, and Rudolf Steiner responds with renewed lecturing vitality. His focus: the spiritual context of human life; initiation science; the course of the year; and community building. As a foundation for an artistic school, he creates a series of pastel sketches. Lecture cycles: *The Anthroposophical Movement; Initiation Science* (CW 227) (in Wales at the Penmaenmawr Summer School); *The Four Seasons and the Archangels* (CW 229); *Harmony of the Creative Word* (CW 230); *The Supersensible Human* (CW 231), given in Holland for the founding of the Dutch society. On November 10, in response to the failed Hitler-Ludendorff putsch in Munich, Steiner closes his Berlin residence and moves the *Philosophisch-Anthroposophisch Verlag* (Press) to Dornach. On December 9, Steiner begins the serialization of his *Autobiography: The Course of My Life* (CW 28) in *Das Goetheanum*. It will continue to appear weekly, without a break, until his death. Late December–early January: Rudolf Steiner re-founds the Anthroposophical Society (about 12,000 members internationally) and takes over its leadership. The new board members are: Marie Steiner, Ita Wegman, Albert Steffen, Elisabeth Vreede, and Gunther Wachsmuth. (See *The Christmas Meeting for the Founding of the General Anthroposophical Society*, CW 260.) Accompanying lectures: *Mystery Knowledge and Mystery Centres* (CW 232); *World History in the Light of Anthroposophy* (CW 233). December 25: the Foundation Stone is laid (in the hearts of members) in the form of the 'Foundation Stone Meditation.'

1924: January 1: having founded the Anthroposophical Society and taken over its leadership, Rudolf Steiner has the task of 'reforming' it. The process begins with a weekly newssheet ('What's Happening in the Anthroposophical Society') in which Rudolf Steiner's 'Letters to Members' and 'Anthroposophical Leading Thoughts' appear (CW 26). The next step is the creation of a new esoteric class, the 'first class' of the 'University of Spiritual Science' (which was to have been followed, had Rudolf Steiner lived longer, by two more advanced classes). Then comes a new language for

Anthroposophy—practical, phenomenological, and direct; and Rudolf Steiner creates the model for the second Goetheanum. He begins the series of extensive 'karma' lectures (CW 235–40); and finally, responding to needs, he creates two new initiatives: biodynamic agriculture and curative education. After the middle of the year, rumours begin to circulate regarding Steiner's health. Lectures: January–February, *Anthroposophy* (CW 234); February: *Tone Eurythmy* (CW 278); June: *The Agriculture Course* (CW 327); June–July: *Speech Eurythmy* (CW 279); *Curative Education* (CW 317); August: (England, 'Second International Summer School'), *Initiation Consciousness: True and False Paths in Spiritual Investigation* (CW 243); September: *Pastoral Medicine* (CW 318). On September 26, for the first time, Rudolf Steiner cancels a lecture. On September 28, he gives his last lecture. On September 29, he withdraws to his studio in the carpenter's shop; now he is definitively ill. Cared for by Ita Wegman, he continues working, however, and writing the weekly installments of his *Autobiography* and *Letters to the Members/Leading Thoughts* (CW 26).

1925: Rudolf Steiner, while continuing to work, continues to weaken. He finishes *Extending Practical Medicine* (CW 27) with Ita Wegman. On March 30, around ten in the morning, Rudolf Steiner dies.

Index

A NOTE FROM RUDOLF STEINER PRESS

We are an independent publisher and registered charity (non-profit organisation) dedicated to making available the work of Rudolf Steiner in English translation. We care a great deal about the content of our books and have hundreds of titles available – as printed books, ebooks and in audio formats.

As a publisher devoted to anthroposophy...

- We continually commission translations of previously unpublished works by Rudolf Steiner and invest in re-translating, editing and improving our editions.
- We are committed to making anthroposophy available to all by publishing introductory books as well as contemporary research.
- Our new print editions and ebooks are carefully checked and proofread for accuracy, and converted into all formats for all platforms.
- Our translations are officially authorised by Rudolf Steiner's estate in Dornach, Switzerland, to whom we pay royalties on sales, thus assisting their critical work.

So, look out for Rudolf Steiner Press as a mark of quality and support us today by buying our books, or contact us should you wish to sponsor specific titles or to support the charity with a gift or legacy.

office@rudolfsteinerpress.com
Join our e-mailing list at www.rudolfsteinerpress.com

RUDOLF STEINER PRESS